List of Substances of the Competent Federal Government and Federal State Authorities

Category "Plants and plant parts"

List of Substances
of the Competent Federal
Government and
Federal State Authorities

Category "Plants and plant parts"

BVL-Reporte

IMPRINT

ISBN 978-3-319-10731-8
ISBN 978-3-319-10732-5 (eBook)
DOI 10.1007/978-3-319-10732-5
Springer Cham Heidelberg New York Dordrecht London

Editor:	Bundesamt für Verbraucherschutz und Lebensmittelsicherheit (BVL) Dienststelle Berlin Mauerstraße 39–42 D-10117 Berlin
Final editing:	Herr K. Bentlage (kb-lektorat), Frau Dr. S. Dombrowski (BVL, Press office)
Editorial board:	Herr F. Riedel (BVL), Herr F. Gründing (LUA Sachsen), Frau R. Maslo (LAVES Niedersachsen), Frau Dr. B. Schlagintweit (LGL Bayern), Frau Dr. R. Schumann (BfR), Frau Dr. K. Stephan (BfArM), Frau K. Woese (LAV Sachsen-Anhalt)
ViSdP:	Frau N. Banspach (BVL, Press office)
Cover design:	deblik, Berlin
Cover illustration:	©Kathleen Rekowski – Fotolia.com
Typesetting:	le-tex publishing services GmbH

Printed on acid-free paper

Springer is part of Springer Science+Business Media (www.springer.com)

Contents

Explanatory notes on the List of Substances of the Competent Federal Government and Federal State Authorities – Category "Plants and plant parts"

1.1 Introduction

The Lists of Substances of the Competent Federal Government and Federal State Authorities (lists of substances) are created to facilitate the classification and assessment of substances regarding their use as food or food ingredient. They are designed to serve as a reference guide for authorities and food distributors. The final assessment of products containing these substances or preparations thereof must always be made on a case-by-case basis, giving due regard to all criteria relevant to that assessment.

Lists of Substances of the Competent Federal Government and Federal State Authorities are to be created for various categories. To provide a uniform system for the lists of substances, the entries in the categories are referred to as "substances", although they may not always be chemically defined individual substances. A more detailed definition may be provided in addition to the categories.

The lists of substances do not claim to be complete and do not exempt the food business operator from the responsibility to ensure that the relevant product is safe and legally marketed as a food in Germany. They are subject to updating in order to take account of new scientific findings as well as developments in the food industry.

1.2 Legal framework

Article 2 of the Regulation (EC) No. 178/2002 (General Food Law Regulation)[1] specifies "food" as any substance or product, whether processed, partially processed or unprocessed, intended to be, or reasonably expected to be ingested by humans.

Substances used as source materials for flavourings or foods with flavouring properties are governed by the provisions of the Regulation (EC) No. 1334/2008 on flavourings and certain food ingredients with flavouring properties for use in and on foods[2].

Foods or food ingredients which have not been used for human consumption to a significant degree in the EU before 15 May 1997 are governed by the provisions of the Regulation (EC) No. 258/97 on novel foods and novel food ingredients (Novel Food Regulation)[3], if they fall into one of the following categories:

- Foods and food ingredients with a new or intentionally modified primary molecular structure,
- Foods and food ingredients consisting of or isolated from micro-organisms, fungi or algae,
- Foods and food ingredients consisting of or isolated from plants and food ingredients isolated from animals, except for foods and food ingredients obtained by traditional propagating or breeding practices and having a history of safe food use
- Foods and food ingredients to which has been applied a production process not currently used, where that process gives rise to significant changes in the composition or structure of the foods or food ingredients which affect their nutritional value, metabolism or level of undesirable substances.

Flavourings for use in foods do not fall within the scope of the Novel Food Regulation. However, if relevant substances are used for purposes other than flavouring, they may need to be classified as novel within the meaning of the Novel Food Regulation.

Article 2 (d) of the General Food Law Regulation stipulates that food does not include medicinal products. The definition of a medicinal product is laid down in Section 2

---

[1] Regulation (EC) No. 178/2002 of the European Parliament and of the Council of 28 January 2002 laying down the general principles and requirements of food law, establishing the European Food Safety Authority and laying down procedures in matters of food safety

[2] Regulation (EC) No. 1334/2008 of the European Parliament and of the Council of 16 December 2008 on flavourings and certain food ingredients with flavouring properties for use in and on foods and amending Council Regulation (EEC) No. 1601/91, Regulations (EC) No. 2232/96 and (EC) No. 110/2008 and Directive 2000/13/EC

[3] Regulation (EC) No. 258/97 of the European Parliament and of the Council of 27 January 1997 concerning novel foods and novel food ingredients

Clause 1 of the Law governing the Trade of Medicinal Products (Medicinal Products Act – The Drug Law, AMG)[4], with which the European definition of medicinal products was enacted into national law in accordance with Article 1 of the Directive 2001/83/EC (Medicinal Products Directive)[5].

Article 2 (g) of the General Food Law Regulation stipulates that food also does not include narcotic and psychotropic substances. According to the General Food Law Regulation, this is laid down in the United Nations Single Convention on Narcotic Drugs of 1961[6] as well as in the United Nations Convention on Psychotropic Substances of 1971[7]. Additionally, the Narcotic Drugs Act (BtMG) applies in German law[8].

1.3 Instruction on how to use the List of Substances

"Substances" within the meaning of this category are plants and plant parts. The substances are viewed and categorised as such. Preparations of substances, such as extracts or isolates, may differ from the actual substances regarding their composition, especially in terms of their nutritional and toxicological properties. In each individual case, it therefore needs to be determined whether the classification of a substance can be applied to a preparation thereof. The classification is made on the basis of a decision tree developed for this category (see Chap. 2).

To understand the List of Substances in the category of "plants and plant parts", the following should additionally be noted:

1. The sub lists have the following meanings:
   - List A: Substances not recommended for use in foods
   - List B: Substances for which restricted use in foods is recommended
   - List C: Substances which cannot yet be completely assessed due to lack of sufficient data
2. If substances are usually used in foods only to a very limited extent, for example as spices or as ingredients in the production of spirits, they are designated by the following abbreviations in the "Food (F)" column:

- F: Known exclusively for use as a food ingredient with flavouring properties or as a source material for flavourings
- S: Known for use as a spice
- C: Known for use as a colouring agent
- T: Known for use as a tea

Such limitations are generally not documented by placing a substance on List B. In exceptional cases, a substance is also placed on List B if it is associated with effects that necessitate a restriction of its use.

Classification is made under the assumption that the relevant substance is used as documented in the list. Any other use, e.g. in higher doses, can lead to effects that may require a different classification.

3. Substances that should be treated before consumption (e.g. heated) are designated by the letter "b" in the "Food (F)" column.
4. When classifying a substance as a novel food/novel food ingredient (NF) within the meaning of the Novel Food Regulation or as not novel in food supplements (Not NFS), the Novel Food Catalogue of the European Commission[9] was taken into account. However, it should be noted that this catalogue does not have any entries for a large number of substances from the List of Substances in the category "plants and plant parts". Any other available information regarding use for human consumption to a significant degree before 15 May 1997 was also taken into account.
5. If any restrictions of use as a food or food ingredient are recommended for a substance (List B) due to evidence of a pharmacological effect (No. 4 in the explanatory notes on the decision tree), such a restriction always refers to the substance described in that evidence (e.g. dried plant or dried plant part). Evidence of a pharmacological effect that leads to a classification as a medicinal product in accordance with Section 2 Clause 1 No. 2a AMG included court-approved sources such as monographs, marketing authorisations or classifications of competent authorities. Other evidence (e.g. results of clinical studies) may additionally be relevant to the classification of a substance as a medicinal product.
6. Substances for which a pharmacological effect or efficacy as a medicinal product is plausible based on use and experience over many years ("traditional evidence") in accordance with Sections 39a ff. of the Medicinal Products Act will not be placed on List B based on that traditional evidence alone. For a substance to be placed on List B, pharmacological effects,

---

[4] Law governing the Trade of Medicinal Products (Medicinal Products Act – AMG)

[5] Directive 2001/83/EC of the European Parliament and of the Council of 6 November 2001 on the Community code relating to medicinal products for human use

[6] Single Convention on Narcotic Drugs of 30 March 1961

[7] Law approving the Convention on Psychotropic Substances of 21 February 1971

[8] Law governing the Trade of Narcotic Drugs (Narcotic Drugs Act – BtMG)

---

[9] http://ec.europa.eu/food/food/biotechnology/novelfood/novel_food_catalogue_en.htm

as specified above in item 5, or risks must be reported due to which restricted use in foods is recommended.

7. The classification of substances into this list is made only on the basis of their effect after oral ingestion. Substances which, according to monographs, are intended for external use only are not designated as medicinal products in the List of Substances.

Decision tree: Explanatory notes on the classification of substances in the List of Substances of the Competent Federal Government and Federal State Authorities – Category "Plants and plant parts"

The decision tree serves as the basis for the classification of substances in the category of "plants and plant parts" as "food (F)", "medicinal product (MP)" and/or "novel food/ novel food ingredient (NF)" as well as possible combinations thereof (ambivalent substances). It also provides instructions for the classification of the substances in the Lists A, B and C.

Classification as a medicinal product – except for ambivalent substances (see No. 3 below) – is made based on the definition of medicinal products by function set out in Section 2 Clause 1 No. 2a of the Medicinal Products Act (AMG). These are characterised by their pharmacological, metabolic or immunological effect. For the sake of clarity, the term "pharmacological effect" is used in the decision tree for this definition. Classification as a medicinal product by presentation as defined in Section 2 Clause 2 No. 1 AMG is not taken into account.

Based on the answers to the questions in the decision tree, plants and plant parts are classified as follows:

No. 1: Food
(Decision tree I – via question 3)
Common foods without any known use as medicinal products. Based on their previous use, any restrictions of use are not required.

No. 2: Food + List B
(Decision tree III – via question 3)
Common foods without any known use as medicinal products.

Dose restrictions and restrictions of use are required due to risks posed by the constituents of the plant or plant part. Such restrictions are expressed by placing the substance on List B.

No. 3: Food + traditional medicinal product
(Decision tree II – via question 5)
For traditional herbal medicinal products, a pharmacological effect is plausible based on use and experience over many years in accordance with Sections 39a ff. of the Medicinal Products Act (AMG). At present, processing monographs for plants/plant parts that can be contained in traditional medicinal products are being created or revised by the European Medicines Agency (EMA). Where these monographs have been adopted and published, they are taken into account accordingly. Some of the plants/plant parts used therein have also long since been used in food. A restriction (List B) was not recommended in individual cases where it would have been done exclusively on the basis of traditional evidence of pharmacological effect. Despite such evidence of pharmacological effect, classification as ambivalent substance (F/MP) without any restrictions is therefore made **in this case only**.

No. 4: Food + medicinal product + List B
(Decision tree III/IV – via question 5)
Common foods that are also used as medicinal products. Pharmacological effects are reported above a certain dose. If no significant pharmacological effects are identified, the substance can be classified as food. When reaching the pharmacologically effective dose, it is defined as a medicinal product by function. This is expressed by placing the substance on List B.

No. 5: Novel food
(Decision tree VIII – via question 9)
The substance is not known to be used as either food or a medicinal product. The substance is also not known to be associated with any risks that would restrict its use in food. The further assessment takes place in accordance with the Novel Food Regulation.

No. 6: Novel food + medicinal product
(Decision tree IV/V – via question 8)
The substance has so far been known as a medicinal product only. However, it is not associated with any risks that would restrict its use in food. Its use in food would there-

List of Substances of the Competent Federal Government and Federal State Authorities, DOI 10.1007/978-3-319-10732-5_2,
© Federal Office of Consumer Protection and Food Safety (BVL) 2014

fore be conceivable after assessment in accordance with the Novel Food Regulation.

Medicinal products and novel foods are distinguished on the basis of the pharmacologically effective dose (by analogy with No. 4). The substance cannot be placed on List B as it is no common food.

No. 7: Novel food (Not NFS) + List C
(Decision tree VIII – via question 9)

A number of substances are classified as not novel exclusively when used in food supplements. When used in foods other than food supplements, these substances are usually placed on List C, since no sufficient information is available for their conclusive assessment.

No. 8: Novel food (Not NFS) + List B + medicinal product
(Decision tree IV/V – via question 8)

The substance is known as a medicinal product. It is additionally used in food supplements and is classified as not novel in this case only. The necessary restriction of use in FS due to the pharmacological effect is expressed by placing the substance on List B.

No. 9: Medicinal product + List A
(Decision tree VI – via question 7)

The substance has so far been known as a medicinal product only. Due to the associated risks, its use in food is not recommended, irrespective of the dose.

No. 10: List A substance
(Decision tree VII – via question 9)

The substance, which is not a medicinal product, is associated with risks. Its use in food is therefore not recommended, irrespective of the dose.

2.1   Classification in the lists A, B and C

List A

Substances that are not recommended for use as food or food ingredient due to known risks (No. 10) are placed on List A. These substances may be or may have been used as medicinal products (No. 9).

List B

Substances that are only used in food and for which a dose restriction is required due to certain constituents (No. 2) are placed on List B.

Furthermore, substances that are known as both foods and medicinal products with a pharmacological effect demonstrated on the basis of clinical data are placed on this list. This is also done by strict application of the decision tree in respect of basic foods in usual amounts of intake (No. 4).

Finally, substances that are used as food exclusively in food supplements (but are otherwise novel foods) and are known as medicinal products with a demonstrable pharmacological effect are placed on list B (No. 8).

Substances that are known as both foods and medicinal products with exclusively traditional evidence of pharmacological effect (No. 3) are not placed on List B.

List C

Substances that have so far been used exclusively in food supplements but are otherwise novel foods are placed on list C (No. 7).

2.2   Decision tree

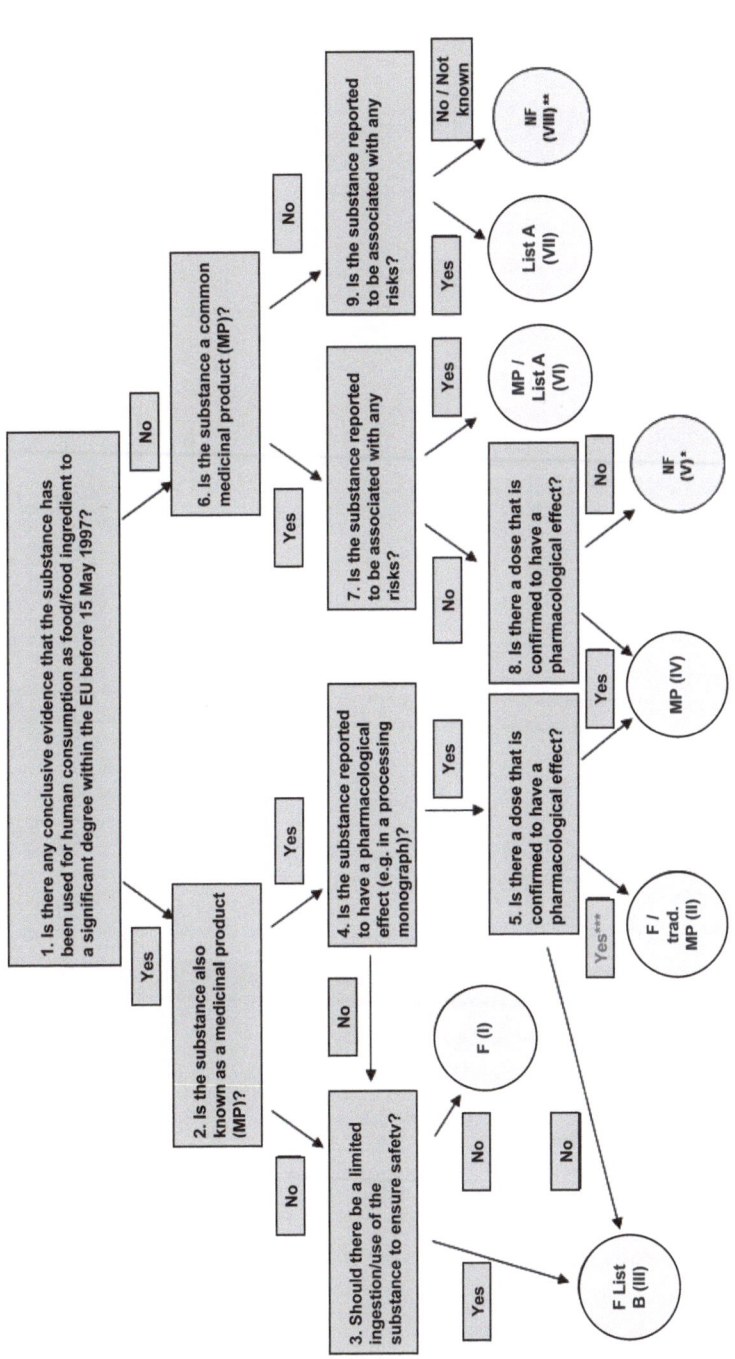

Figure 1   Decision tree
* Assessment of potential use as food/food ingredient according to the Novel Food Regulation (If the substance is classified as not novel in food supplements [Not NFS], it may need to be placed on List B.)
** Assessment of potential use as food/food ingredient according to the Novel Food Regulation (If the substance is classified as not novel in food supplements [Not NFS], it may need to be placed on List C.)
*** See No. 3 of the explanatory notes
Abbreviations: F = food, MP = medicinal product, trad. MP = traditional medicinal product, NF = novel food/novel food ingredient, FS = food supplement

List of Substances of the Competent Federal Government and Federal State Authorities – Category "Plants and plant parts"

Status: December 2013

List of Substances of the Competent Federal Government and Federal State Authorities, DOI 10.1007/978-3-319-10732-5_3,
© Federal Office of Consumer Protection and Food Safety (BVL) 2014

Table 1   List of Substances – Category "Plants and plant parts"

Stock plant (Latin)	Plant part	F	NF	MP	trad. MP	List A	List B	List C
Abies spp.	Shoot (tip)	× T						
Achillea millefolium L.	Herb, flower	× S, T		×	×		×	
Aconitum napellus L.	all plant parts			×	×			
Acorus calamus L., syn. Acorus aromaticus GARZ., Oronthium cochinchinensis LOUR.	Rhizome	× F			×		×	
Actinidia deliciosa (CHEV.) A.R. FERG.	Fruit	×						
Adonis vernalis L., syn. Adonanthe vernalis SPACH	Herb			×	×			
Aegopodium podagraria L., syn. Aegopodium angelicaefolium ST. LAGER, Aegopodium latifolium TURCZ.	Leaf	×						
Aesculus hippocastanum L., syn. Aesculus castanea GILIB., Aesculus procera SALISB., Hippocastanum vulgare GAERTNER	Leaf, flower		Not NFS					×
Aesculus hippocastanum L., syn. Aesculus castanea GILIB., Aesculus procera SALISB., Hippocastanum vulgare GAERTNER	Bark		Not NFS		×			×

Remarks	Risks	Critical plant substances	Pharmacologically effective dose
Positive monograph, indications: loss of appetite, digestive problems such as mild gastrointestinal cramps, **standard marketing authorisation:** 1249.99.99, **ESCOP monograph:** loss of appetite, dyspeptic disorders such as mild spasmodic complaints in the abdominal region, **WHO monograph:** loss of appetite, common cold, dyspeptic ailments such as mild spastic discomfort of the gastrointestinal tract, as a choleretic and for the treatment of fevers, **HMPC monograph:** traditional use	Contraindications: hypersensitivity to yarrow or other Asteraceae species (Commission E monograph)	Proazulenes, azulenes, monoterpenes (e.g. camphor, eucalyptol [=1,8-cineol])	**Commission E:** 4 g drug/day, equivalent preparations **ESCOP:** 2–4 g drug/3–4 times a day, equivalent preparations, administration to children aged between 3 and 12 only under medical supervision **WHO:** 4.5 g drug/day or 3 g flower tea/day
Negative monograph: due to risks, use in MP only in homeopathic doses, monkshood (Aconitum napellus), extremely toxic Ia (Wink/Wyk), extremely toxic +++ (Roth/Daunderer)	Aconitine is a strong nerve and muscle poison, numbness, paraesthesia, paralysis (Wink/Wyk), cardiac arrhythmia, sensitivity to cold, nausea, cramps, paralysis of tongue, facial and extremity muscles, circulatory failure (Roth/Daunderer)	Alkaloids (e.g. aconitine)	
Risks reported (in dependence on the variant used), use of tetraploid subspecies as source material for flavourings and foods with flavouring properties prohibited by Reg. (EC) No. 1334/2008, maximum amount of beta-asarone in alcoholic beverages: 1 mg/kg, licensed anthroposophic medicinal product acc. to Section 21 AMG, toxic + (Roth/Daunderer), mildly toxic II to toxic III (Wink/Wyk)	Risks reported for tetraploid variants e.g. A. calamus var. angustatus BESS. because of their high beta-asarone content. Beta-asarone demonstrated a mutagenic and carcinogenic effect in animal experiments (Hager, Frohne/Pfänder); diploid subspecies e.g. A. calamus L. var. americanus (RAF.) WULFF. are free from beta-asarone	Beta-asarone	
Positive monograph, indications: slightly reduced cardiac output, especially when accompanied by nervous symptoms, very toxic Ib (Wink/Wyk), highly toxic ++ (Roth/Daunderer)	Nausea, vomiting, cardiac arrhythmia, contraindications and interactions reported (Commission E monograph), cardiac arrest, symptoms of cardiac glycoside poisoning, diuresis, digestive tract irritation (Wink/Wyk)	cardioactive glycosides (some prescription-only)	**Commission E:** 0.6–3 g standardised Adonis powder (DAB 9)/day, equivalent preparations
Neutral monograph			
Neutral monograph, HMPC monograph: traditional use			

Stock plant (Latin)	Plant part	F	NF	MP	trad. MP	List A	List B	List C
Aesculus hippocastanum L., syn. Aesculus castanea GILIB., Aesculus procera SALISB., Hippocastanum vulgare GAERTNER	Seed		Not NFS	×	×		×	
Agrimonia eupatoria L., syn. Agrimonia adherens GILIB., Agrimonia officinalis LAM., Agrimonia parviflora SPRENG., Amonia agrimonoides STEUD., Aremonia agrimonoides D.C.	Herb	× T		×			×	
Agrimonia procera WALLR., syn. Agrimonia odorata auct. non MILLER, Agrimonia odorata WALLR.	Herb	× T		×			×	
Agropyron repens (L.) P. BEAUV., syn. Elymus repens (L.) GOULD, Elytrigia repens DESV. ex NEVSKI, Triticum repens L.	Root	× T		×	×		×	
Alchemilla alpina L. em. BUSER	Herb		×		×			
Alchemilla vulgaris auct., Alchemilla vulgaris L.	Herb	× T		×			×	
Alkanna tuberculata (FORSSK.) MEIKLE, syn. Alkanna tinctoria (L.) TAUSCH, Alkanna tuberculata GREUTER, Lithospermum tinctorium L.	Root					×		
Allium cepa L., syn. Allium esculentum SALISB., Cepa esculenta S.F. GRAY, Cepa vulgaris RENAULT, Kepa esculenta RAFIN., Porrum cepa RCHB.	Bulb	×		×			×	

Remarks	Risks	Critical plant substances	Pharmacologically effective dose
Positive monograph, indications: symptomatic treatment of diseases of crural veins (chronic venous insufficiency), e.g. pain and feeling of heaviness in legs, night leg cramps, itching and swollen legs, **WHO monograph:** treatment of symptoms of chronic venous insufficiency, including pain, feeling of heaviness in the legs, nocturnal calf-muscle spasms, itching and oedema, **ESCOP monograph:** chronic venous insufficiency, varicosis, **HMPC monograph:** treatment of chronic venous insufficiency, which is characterised by swollen legs, varicose veins, a feeling of heaviness, pain, tiredness, itching, tension and cramps in the calves, traditional use, slightly toxic + (Roth/Daunderer), mildly toxic III (Wink/Wyk)	Nausea, gastrointestinal complaints, hot flushes, oedemas, vomiting, diarrhoea, hypertension, loss of consciousness, circulatory collapse (Wink/Wyk, Roth/Daunderer)	Saponins	**Commission E:** 100 mg aescin/day, equivalent to 250–312.5 mg extract in delayed release form 2 times a day **ESCOP:** equivalent to 50–150 mg triterpene glycosides (calculated as aescin), no administration to children **WHO:** 250–312.5 mg standardised comminuted extract, equivalent to 100 mg aescin
Positive monograph, indications: mild, non-specific, acute forms of diarrhoea, inflammations of oral and pharyngeal mucosa, **ESCOP monograph:** Agrimony has widely documented uses as a remedy to treat mild diarrhoea, **HMPC monograph:** in progress, **standard marketing authorisation:** 2379.99.99			**Commission E:** 3–6 g drug/day, equivalent preparations **ESCOP:** 3–12 g drug, 3 times a day, children!
Positive monograph, indications: mild, non-specific, acute forms of diarrhoea, inflammations of the oral and pharyngeal mucosa, **HMPC monograph:** in progress			**Commission E:** 3–6 g drug/day, equivalent preparations
Positive monograph, indications: increase of urinary output for catarrhs of the lower urinary tract, adjuvant treatment of catarrhs of the upper respiratory tract, **standard marketing authorisation:** 1169.99.99, contraindications reported, **ESCOP monograph:** irritable bladder and other urinary tract disorders, **HMPC monograph:** traditional use	Contraindications: irrigation therapy of oedemas associated with cardiac or renal insufficiency (Commission E monograph)		**Commission E:** 6–9 g drug/day, equivalent preparations **ESCOP:** 5–10 g drug/day, no administration to children under the age of 12
Neutral monograph			
Positive monograph, indications: mild, non-specific forms of diarrhoea			**Commission E:** 5–10 g drug/day, equivalent preparations
Substance of concern (AMK – Drug Commission of German Pharmacists), BfArM [Federal Institute for Drugs and Medical Devices] graduated plan of 5 June 1992 concerning medicinal products containing pyrrolizidine alkaloids, toxic II (Wink/Wyk), toxic + (Roth/Daunderer)	Pyrrolizidine alkaloids contained in the drug are mutagenic, carcinogenic, ingestion of high doses causes digestive problems and CNS disorders (Wink/Wyk)	Pyrrolizidine alkaloids	
Positive monograph, indications: loss of appetite, prevention of age-dependent vascular changes, **WHO monograph:** the principal use of Bulbus Allii Cepae today is to prevent age-dependent changes in the blood vessels, and loss of appetite, non-toxic but irritating to skin (Roth/Daunderer), mildly toxic III (Wink/Wyk), **HMPC assessment report**	Ingestion of higher doses, especially in children, causes digestive tract irritation accompanied by vomiting, nausea, colic and diarrhoea (Wink/Wyk)	Allicin, allyl sulphide	**Commission E:** 50 g fresh onions or 20 g dried drug, equivalent preparations **WHO:** 20 g drug/day, equivalent preparations

Stock plant (Latin)	Plant part	F	NF	MP	trad. MP	List A	List B	List C
Allium porrum L., syn. Allium ampeloprasum ß. porrum GAY, Allium laetum SALISB.	Leaf	×						
Allium sativum L., syn. Porrum sativum RCHB.	Bulb	×		×			×	
Allium ursinum L., syn. Allium latifolium GILIB., Allium nemorale SALISB.	Herb	× S			×			
Aloe barbadensis MILL., Aloe chinensis BAK.	Leaf gel	×						
Aloe barbadensis MILL., Aloe chinensis BAK.	Leaf juice			×		×		

Remarks	Risks	Critical plant substances	Pharmacologically effective dose
Positive monograph, indications: adjuvant treatment of increased blood lipid concentrations, prevention of age-related vascular changes, **WHO monograph:** as an adjuvant to dietetic management in the treatment of hyperlipidaemia, and in the prevention of atherosclerotic (age-dependent) vascular changes. The drug may be useful in the treatment of mild hypertension, **ESCOP monograph:** prophylaxis of atherosclerosis, treatment of elevated blood lipid levels insufficiently influenced by diet, also used for respiratory tract infections and catarrhal conditions, although clinical data to support this indication is not available, **HMPC monograph:** in progress, BfArM graduated plan of 11 Mar 2008 concerning medicinal products containing plant substances or preparations of Allium sativum (http://www.bfarm.de/DE/Pharmakovigilanz/risikoinfo/stufenplanverf/Liste/stp-allium-sativum.html), toxic for humans only after overdose + (Roth/Daunderer), mildly toxic III (Wink/Wyk)	Higher doses may have a toxic effect (especially in children), digestive tract irritation accompanied by vomiting, excessive sweating, nausea, colic and diarrhoea (Wink/Wyk)	Alliin, allicin, allyl sulphide	**Commission E:** 4 g fresh garlic/day, equivalent preparations **ESCOP:** 0.5–1 g drug/day **WHO:** 0.4–1.2 g powder or 2–5 g fresh garlic/day
non-toxic or hardly toxic (+) (Roth/Daunderer), mildly toxic III (Wink/Wyk)	Ingestion of higher doses, especially in children, causes digestive tract irritation accompanied by vomiting, nausea, colic and diarrhoea (Wink/Wyk)	Allicin, allyl sulphide	
Mucilage from the parenchymatous tissue of the leaf interior, free from anthranoids, addition of isolated aloin prohibited by Reg. (EC) No. 1334/2008, **WHO monograph:** (traditional use, external)			**WHO:** 10–70 % of fresh gel for external use
Positive monograph, indications: constipation, **WHO monograph:** short-term treatment of occasional constipation, **ESCOP monograph:** for short term use in cases of occasional constipation, **HMPC monograph:** short-term use in cases of occasional constipation, BfArM graduated plan of 21 June 1996 concerning medicinal products for human use containing anthranoid-containing plant genera, toxic II (Wink/Wyk), toxic + (Roth/Daunderer)	gastrointestinal cramps observed in individual cases, long-term use/abuse causes electrolyte loss, in particular potassium, albuminuria and haematuria. Loss of potassium may cause cardiac dysfunction and muscular asthenia (Commission E processing monograph), menstrual and uterine haemorrhage. Long-term use observed to cause functional disorders of the genital system accompanied by follicular atresia and hypertrophy of interstitial tissue. Abortion may occur, strong laxative effect, intestinal bleeding, tenesmus, haemorrhoids, increased diuresis, renal irritation (Roth/Daunderer)	Anthranoids (e.g. aloin/Aloe emodin)	**Commission E, ESCOP:** equivalent to 10–30 mg hydroxyanthracene derivatives (calculated as barbaloin)/day, no administration to children under the age of 10 **WHO:** 0.1 g drug/day, single dose: 40–110 mg **HMPC:** equivalent to 10–30 mg hydroxyanthracene derivatives (calculated as barbaloin)/day, no administration to children under the age on 12

Stock plant (Latin)	Plant part	F	NF	MP	trad. MP	List A	List B	List C
Aloe ferox MILL., syn. Aloe horrida HAW., Pachidendron ferox HAW.	Leaf juice			×		×		
Alpinia galanga (L.) WILLD., syn. Maranta galanga L.	Rhizome	× S, T						
Alpinia officinarum HANCE	Rhizome	× S, T		×			×	
Althaea officinalis L.,	Leaf	× T		×			×	
Althaea officinalis L.,	Root	× T		×	×		×	
Althaea rosea L.	Flower	× T						
Amaranthus caudatus L.	Seed	×						
Amaranthus cruentus L.	Seed	×						
Amaranthus tricolor L.	Herb	×						
Ammi visnaga (L.) LAM., syn. Daucus visnaga L., Apium visnaga CRANTZ	Fruit		Not NFS	×			×	
Ananas comosus (L.) MERR.	Fruit	×						
Anemone spp.	all plant parts					×		
Anemopaegma mirandum MART.	Bark		Not NFS					×
Anethum graveolens L.	Fruit	× S		×	×		×	
Anethum graveolens L.	Herb	× S						
Angelica archangelica L., syn. Angelica archangelica ALL.	Stem	× T, S						

Remarks	Risks	Critical plant substances	Pharmacologically effective dose
Positive monograph, indications: constipation, **WHO monograph:** short-term treatment of occasional constipation, **ESCOP monograph:** for short term use in cases of occasional constipation, **HMPC monograph:** short-term use in cases of occasional constipation, BfArM graduated plan of 21 June 1996 concerning medicinal products for human use containing anthranoid-containing plant genera, toxic II (Wink/Wyk), toxic (Roth/Daunderer)	gastrointestinal cramps observed in individual cases, strong laxative effect, intestinal bleeding, tenesmus, haemorrhoids, increased diuresis, electrolyte loss (in particular potassium), which may cause cardiac dysfunction and muscular asthenia, albuminuria, haematuria (Commission E monograph), menstrual and uterine haemorrhage. Long-term use observed to cause functional disorders of the genital system accompanied by follicular atresia and hypertrophy of interstitial tissue, abortion may occur (Roth/Daunderer)	Anthranoids	**Commission E:** 20–30 mg hydroxyanthracene derivatives/ day, calculated as anhydrous aloin **ESCOP:** preparations equivalent to 10–30 mg hydroxyanthracene derivatives (calculated as barbaloin)/day, no administration to children under the age of 10 **HMPC:** equivalent to 10–30 mg hydroxyanthracene derivatives (calculated as barbaloin)/day, no administration to children under the age of 12
Positive monograph, indications: digestive problems, loss of appetite			**Commission E:** 2–4 g drug/day, equivalent preparations
Positive monograph, indications: inflammations of the oral and pharyngeal mucosa, dry cough, mild inflammations of the gastric mucosa, **standard marketing authorisation:** 1469.99.99			**Commission E:** 5 g drug/day, equivalent preparations
Positive monograph, indications: inflammations of the oral and pharyngeal mucosa, dry cough, **ESCOP monograph:** dry cough, irritation of the oral pharyngeal or gastric mucosa, **WHO monograph, HMPC monograph:** traditional use, **standard marketing authorisation:** 8899.99.99			**Commission E:** 6 g drug/day, equivalent preparations **ESCOP, WHO:** 0.5–3 g drug, up to 15 g drug/day
Negative monograph: due to risks, **WHO monograph:** as anantispasmodic, muscle relaxant and vasodilator, non-toxic or hardly toxic (+) (Roth/Daunderer)	slightly phototoxic, overdose and long-term use cause vertigo, nausea, constipation, lack of appetite, headache (Hänsel/Sticher, Roth/Daunderer)	Furanochromones (e.g. visnagin, khellin)	**WHO:** 50–150 mg drug/day, administration to children only under medical supervision
toxic II (Wink/Wyk), toxic + (Roth/Daunderer)	Nausea and CNS disorders, digestive and kidney problems, blistering and ulceration, skin rash (Wink/Wyk)	Protoanemonin	
individually or mixed with other barks designated as catuaba			
Positive monograph, indications: digestive problems, **WHO monograph:** treatment of dyspepsia, gastritis and flatulence, and stomach ache			**Commission E, WHO:** 3 g drug/ day, equivalent preparations
Neutral monograph			
Negative monograph: due to risks, slightly toxic (+) (Roth/Daunderer)	phototoxic	Coumarins, furocoumarins	

Stock plant (Latin)	Plant part	F	NF	MP	trad. MP	List A	List B	List C
Angelica archangelica L., syn. Angelica archangelica ALL.	Root	× F		×	×		×	
Angelica silvestris L., syn. Angelica brachyradia FREYN	all plant parts		Not NFS					×
Angelica sinensis (OLIV.) DIELS	all plant parts		Not NFS	×				×
Antennaria dioica (L.) GAERTN., syn. Cyttarium dioicum (L.) PETERM., Gnaphalium dioicum L.	Flower	× C						
Anthriscus cerefolium (L.) HOFFM. var. cerefolium [var. sativus (LAM.) ENDL.]	Herb	× S						
Apium graveolens L., syn. Apium celleri GÄRTNER, Celeri graveolens BRITTON, Selinum graveolens E.H.L. KRAUSE, Seseli graveolens SCOP., Sison ruta BURM.	Leaf	× S						
Apium graveolens L., syn. Apium celleri GÄRTNER, Celeri graveolens BRITTON, Selinum graveolens E.H.L. KRAUSE, Seseli graveolens SCOP., Sison ruta BURM.	Fruit	× S, T						
Apium graveolens L., syn. Apium celleri GÄRTNER, Celeri graveolens BRITTON, Selinum graveolens E.H.L. KRAUSE, Seseli graveolens SCOP., Sison ruta BURM.	Root, stem	×						
Aquilegia vulgaris L., syn. Aquilegia cornuta GILIB.	all plant parts					×		
Arctium lappa L., syn. Articum majus BERNH., Lappa major GAERTN.	Root	× T			×			
Arctostaphylos uva-ursi (L.) SPRENGEL, syn. Arbutus uva-ursi L., Arctostaphylos media GREENE	Leaf		Not NFS	×			×	

Remarks	Risks	Critical plant substances	Pharmacologically effective dose
Positive monograph, indications: lack of appetite, digestive problems such as mild gastrointestinal cramps, feeling of fullness, bloating, **standard marketing authorisation:** 1419.99.99, **ESCOP monograph:** dyspeptic complaints such as mild gastrointestinal spasms, sluggish digestion, flatulence and feeling of fullness, loss of appetite, anorexia, bronchitis, slightly toxic (+) (Roth/Daunderer)	phototoxic	Coumarins, furo-coumarins	**Commission E:** daily dose: 4.5 g drug or 1.5–3 g fluid extract (1:1) or 1.5 g tincture (1:5), equivalent preparations **ESCOP:** 3–6 g drug/day, no administration to children
WHO monograph (root): traditional use, **HMPC assessment report**	phototoxic	Coumarins, furo-coumarins	**WHO:** 4.5–9 g drug/day, no administration to children
Neutral monograph			
		Estragole (=methyl chavicol)	
Negative monograph: due to risks	phototoxic, Commission E: celery may induce allergic reactions and even anaphylactic shock (celery-carrot-mugwort syndrome)	Furocoumarins	
Negative monograph: due to risks	phototoxic, Commission E: celery may induce allergic reactions and even anaphylactic shock (celery-carrot-mugwort syndrome)	Furocoumarins	
Negative monograph: due to risks	phototoxic, Commission E: celery may induce allergic reactions and even anaphylactic shock (celery-carrot-mugwort syndrome)	Furocoumarins	
toxic II (Wink/Wyk), toxic + (Roth/Daunderer), whole plant, especially the seeds	Vertigo, pupillary changes, dizziness, loss of consciousness, dyspnoea (Wink/Wyk), diarrhoea (Roth/Daunderer)	Isoquino-line alkaloids (e.g. mag-noflorine), cyanogenic glyco-sides	
HMPC monograph: traditional use			
Positive monograph, indications: inflammatory diseases of the lower urinary tract, **WHO monograph, HMPC monograph:** traditional use, **standard marketing authorisation:** 8299.99.99, **ESCOP monograph:** uncomplicated infections of the lower urinary tract such as cystitis, when antibiotic treatment is not considered essential, mildly toxic III (Wink/Wyk), slightly toxic (+) (Roth/Daunderer)	Nausea and vomiting may occur in people with a weak stomach (Commission E monograph), uterus contractions, long-term use may cause symptoms of poisoning such as haemolytic anaemia, cachexia, fatty liver disease and hair discolouration resulting from the release of hydroquinones and/or methylhydroquinones in the body, contraindications and interactions reported (Roth/Daunderer)	Hydroquinones (e.g. arbutin)	**Commission E, WHO:** 3 g in 150 ml water, up to 4 times a day, no administration to children under the age of 12 **ESCOP:** equivalent to 400–800 mg arbutin/day, no administration to children

Stock plant (Latin)	Plant part	F	NF	MP	trad. MP	List A	List B	List C
Arctostaphylos uva-ursi (L.) SPRENGEL, syn. Arbutus uva-ursi L., Arctostaphylos media GREENE	Fruit		×					
Areca catechu L.	Fruit					×		
Argyreia nervosa (BURM.f.) BOJ., syn. Argyreia speciosa (L.f.) SWEET, Convolvulus nervosus BURM.f.	Seed					×		
Aristolochia spp.	all plant parts			×		×		
Armoracia lapathifolia GILIB., syn. Armoracia rusticana G.M. SCH., Armoracia sativa BERNH., Cardamine armoracia O. KTZE.	Root	×		×			×	
Arnica montana L., syn. Doronicum arnica DESF.	Flower					×		
Aronia melanocarpa (MICHX.) ELLIOT	Fruit	×						
Artemisia abrotanum L., syn. Artemisia abrotanifolium SALISB., Artemisia angustifolia S.F. GRAY	Herb	× S, T						
Artemisia absinthium L., syn. Absinthium majus GEOFFR., Absinthium officinale LAM., Absinthium vulgare LAM.	Herb	× T		×	×		×	

Remarks	Risks	Critical plant substances	Pharmacologically effective dose
toxic II (Wink/Wyk), toxic + (Roth/Daunderer), psychotropic. Consumption in the form of "betel quids" consisting of areca seeds, betel leaves (Piper betle), dissolved lime as well as flavour-enhancing additives. When chewed, the alkaloid arecaidine (hydrolysis of arecoline) is released, which has a central stimulating effect. CNS-stimulant, narcotic	neurotoxic, arecoline activates mAChR, therefore psychotropic, stimulates salivation, bradycardia, respiratory and cardiac arrest. People who regularly chew betel quids often suffer from oral tumours (Wink/Wyk). Higher doses (8–10 g seeds) may be fatal (Hager)	Alkaloids (e.g. arecoline, arecaidine)	
toxic Ib to very toxic II (Wink/Wyk)	psychotropic, highly psychedelic, effect similar to that of LSD, seeing colours (Wink/Wyk)	Ergot alkaloids (e.g. lysergic acid, ergine = lysergic acid amide, LSA)	
BfArM graduated plan of 3 June 1981 concerning medicinal products containing aristolochic acid, including phytotherapeutic and homeopathic medicinal products, and BfArM graduated plan of 22 June 2010 concerning medicinal products produced using plants of the Asarum genus of the Aristolochiaceae family, substance of concern (AMK), toxic II (Wink/Wyk), toxic + (Roth/Daunderer)	Vomiting, gastroenteritis, drop in blood pressure, cramps, tachycardia, serious damage to urinary tract and kidneys may occur, aristolochic acids have a mutagenic and carcinogenic effect, ingestion of higher doses may cause death in coma by apnoea (BfArM, Roth/Daunderer)	Aristolochic acids	
Positive monograph, indications: catarrhs of the respiratory tract, adjuvant treatment of lower urinary tract infections	gastrointestinal complaints, contraindications reported (Commission E monograph)		**Commission E:** 20 g fresh root/day, equivalent preparations
Positive monograph for external use only, **standard marketing authorisation:** 5799.99.99 (tincture), 8199.99.99 (flower), **WHO monograph, ESCOP monograph:** for external use only, toxic II (Wink/Wyk), slightly toxic (+) but often sensitising (Roth/Daunderer), **HMPC monograph:** in progress (external use)	Skin irritation, severe gastroenteritis accompanied by diarrhoea and gastric pain, headache, vertigo, bradycardia, anxiety and agitation, strong heartbeat, abortifacient effect (Wink/Wyk), due to its toxic side effects observed, arnica should not be used internally (Roth/Daunderer) (AMKdA)	Sesquiterpene lactones (e.g. helenalin)	**Commission E, ESCOP, WHO:** 2 g drug/day for external use
Positive monograph, indications: loss of appetite, digestive problems, dyskinesia of the biliary tract, **ESCOP monograph:** Anorexia, for example after illness, dyspeptic complaints, **HMPC monograph:** traditional use, **standard marketing authorisation:** 1339.99.99, mildly toxic II to toxic III (Wink/Wyk), slightly toxic (+) to toxic + (Roth/Daunderer)	Vomiting, severe diarrhoea, dizziness, headache and cramps (Wink/Wyk)	Thujone	**Commission E:** 2–3 g drug/day as water extract **ESCOP:** 1–1.5 g drug in 150 ml water up to 3 times a day

Stock plant (Latin)	Plant part	F	NF	MP	trad. MP	List A	List B	List C
Artemisia cina O.C. BERG et C.F. SCHMIDT, syn. Artemisia cina BERG, Artemisia mogoltavica POLJAK.	Flower, seed					×		
Artemisia dracunculus L., syn. Artemisia inodora WILLD., Artemisia redowskyi LEBED., Oligosporus condimentarius CASS.	Leaf	× T, S						
Artemisia vulgaris L., syn. Artemisia lactiflora WALL. ex DC., Artemisia samamisica BESS.	Herb	× S, T			×			
Arum spp.	all plant parts					×		
Asarum europaeum L., syn. Asarum officinale MOENCH.	all plant parts					×		
Aspalathus linearis (BURM.F.) DAHLG., syn. Aspalathus contaminata (THUNB.) DRUCE, Aspalathus cognata PRESL., Borbonia pinifolia MARL., Lebeckia linearis (BURM.F.) DC., Psoralea linearis BURM.F.	Herb	× T						
Asparagus officinalis L., syn. Asparagus altilis ASCHERS., Asparagus hortensis MILL.	Rhizome	×		×			×	
Aspidosperma quebracho-blanco SCHLECHT., syn. Aspidosperma chakensis SPEG., Macaglia quebracho O. KTZE.	Bark, wood					×		
Astragalus membranaceus (FISCH) BGE., syn. Phaca membranacea FISCH. INDC.	Root		Not NFS		×			×

Remarks	Risks	Critical plant substances	Pharmacologically effective dose
very toxic Ib (Wink/Wyk), highly toxic ++ (Roth/Daunderer)	Loss of consciousness, epileptic fits, visual disturbances, excessive gall production, icterus, diarrhoea, nausea, vomiting, cold spells, hypersalivation, excessive lacrimation, dyspnoea, facial muscle cramps, haematuria resulting from kidney damage, crural paralysis, delirium, coma, death by respiratory arrest (Wink/Wyk)	Sesquiterpene lactones (e.g. santonin, artemisin)	
concerning estragole content, s. BfR statement: http://www.bfr.bund.de/cm/208/minimierung_von_estragol_und_methyleugenol_gehalten_in_lebensmitteln.pdf	Estragole and methyl eugenol demonstrated a genotoxic effect in in-vitro and in-vivo animal experiments (SCF, BfR)	Estragole (=methyl chavicol), methyl eugenol	
Negative monograph: due to risks	abortifacient effect reported, allergic reactions may be induced after previous sensitisation (Commission E monograph), belongs to the most potent plant allergens (Schilcher)	Thujone, camphor, eucalyptol (=1,8-cineol)	
extremely toxic Ia (Wink/Wyk), extremely toxic +++ (Roth/Daunderer)	Irritating to skin, burning in the mouth and throat, cardiac arrhythmia, cramps, internal haemorrhage, CNS disorders, digestive problems (Wink/Wyk), hypersalivation, gingival, gastrointestinal and uterine bleeding (Roth/Daunderer)	Aroin, saponins, cyanogenic glycosides	
toxic II (Wink/Wyk), toxic + (Roth/Daunderer), substance of concern (AMK), due to detection of aristolochic acids, BfArM graduated plan procedure resulted in recall of medicinal products containing Asarum species: http://www.bfarm.de/cae/servlet/contentblob/1207042/publicationFile/	Ingestion of high doses causes digestive problems and CNS disorders, burning in the mouth and throat, nausea, vomiting, gastric pain, gastroenteritis accompanied by diarrhoea and uterine haemorrhage, in extreme cases death by central apnoea (Wink/Wyk, Roth/Daunderer)	Asarone, aristolochic acids	
Positive monograph, indications: irrigation therapy of inflammatory diseases of the lower urinary tract and prevention of kidney gravel			**Commission E:** 45–60 g drug, equivalent preparations
highly toxic ++ (Roth/Daunderer), toxic Ib to very toxic II (Wink/Wyk)	affects vessels (vasoconstrictive), affects respiratory activity, local anaesthetic action, emetic in high doses (Roth/Daunderer, Wink/Wyk)	Indole alkaloids (e.g. aspidospermine, yohimbine)	
WHO monograph: traditional use			**WHO:** 9–30 g/day

Stock plant (Latin)	Plant part	F	NF	MP	trad. MP	List A	List B	List C
Atropa belladonna L.	all plant parts			×		×		
Avena sativa L., syn. Avena cinerea ROEM. et SCHULT.	Herb	×			×			
Avena sativa L., syn. Avena cinerea ROEM. et SCHULT.	Fruit	×			×			
Backhousia citriodora F. MUELL.	Leaf	× S						
Banisteriopsis caapi (SPRUCE ex GRISEB.) MORTON, syn. Banisteriopsis inebrians MORTON	Bark, wood					×		
Barosma betulina (THUNB.) BARTL. et WENDL., syn. Agathosma betulina (BERG.) PILLANS	Leaf	× F			×			
Bellis perennis L.	Flower	× T						
Berberis vulgaris L.	Fruit	×						
Berberis vulgaris L.	Bark			×		×		
Berberis vulgaris L.	Root, root cortex					×		
Beta vulgaris L. ssp. vulgaris var. Conditiva ALEF.	Tuber	×						
Betula pendula ROTH and Betula pubescens EHRH.	Leaf	× T		×	×		×	

Remarks	Risks	Critical plant substances	Pharmacologically effective dose
Positive monograph: cramps and colic-like pain in the gastrointestinal and biliary tract, extremely toxic Ia (Wink/Wyk), extremely toxic +++ (Roth/Daunderer), isolated tropane alkaloids contained in the drug prescription-only, deadly nightshade pharmacy-only	psychotropic, all plant parts extremely toxic, dry mouth, accommodation difficulties, reddened and dry skin, hyperthermia, tachycardia, hallucinations, cramps, micturition difficulties, delirium, fever, cardiac arrest, apnoea, death, contraindications and interactions with medicinal products reported (Frohne, Commission E monograph)	Tropane alkaloids (e.g. atropine, L-hyoscyamine, scopolamine)	**Commission E:** Belladonnae pulvis normatus: average single dose 0.05–0.1 g, max. single dose: 0.2 g equivalent to 0.6 mg total alkaloids, max. daily dose: 0.6 g equivalent to 1.8 mg total alkaloids; Belladonnae radix: average single dose: 0.05 g, max. daily dose: 0.3 g equivalent to 1.5 mg total alkaloids
Neutral monograph, HMPC monograph: traditional use			
Neutral monograph, HMPC monograph: traditional use			
highly toxic ++ (Roth/Daunderer), toxic II (Wink/Wyk), used to produce ayahuasca, yagé or caapi	Beta-carboline alkaloids contained in the drug are strong monoamine oxidase (MAO) inhibitors, hallucinations, mydriasis, hypersalivation, vomiting, sweating, nausea, tremor, abdominal pain, rectal tenesmus (Wink/Wyk)	Beta-carboline alkaloids (e.g. harmala, harmaline)	
Neutral monograph	gastric irritation (Schilcher)	Diosphenol and pulegone	
Neutral monograph: toxic II (Wink/Wyk), slightly toxic (+) (Roth/Daunderer). Ripe berries are edible (Roth/Daunderer), alkaloid content decreases with increasing maturity (Hager)		Isoquino-line alkaloids (e.g. berberine)	
Negative monograph: due to risks, toxic II (Wink/Wyk), slightly toxic (+) (Roth/Daunderer), **WHO monograph:** used orally for the treatment of digestive complaints, such as dyspepsia, diarrhoea, gastritis, feelings of distention and flatulence	Dizziness, nasal bleeding, vomiting, diarrhoea, renal irritation, nephritis, ingestion of higher doses causes poisoning by respiratory arrest and haemorrhagic nephritis (Wink/Wyk)	Isoquino-line alkaloids (e.g. berberine)	**WHO:** 0.5–1 g drug, 3 times a day, no administration to children under the age of 12
Negative monograph: due to risks, toxic II (Wink/Wyk), slightly toxic (+) (Roth/Daunderer)	Dizziness, nasal bleeding, vomiting, diarrhoea, renal irritation, nephritis, ingestion of higher doses causes poisoning by respiratory arrest and haemorrhagic nephritis (Wink/Wyk)	Isoquino-line alkaloids (e.g. berberine)	
Positive monograph, indications: irrigation therapy of bacterial, inflammatory and other diseases of the lower urinary tract and kidney gravel, adjuvant treatment of rheumatic symptoms, **standard marketing authorisation:** 8399.99.99, **ESCOP monograph:** irrigation of the urinary tract, especially in cases of inflammation and renal gravel and as an adjuvant in the treatment of bacterial infections, **HMPC monograph:** traditional use	Contraindications: irrigation therapy of oedemas associated with cardiac or renal insufficiency (Commission E monograph)		**Commission E:** 6–10 g drug/day, equivalent preparations, average daily dose: 2.0–3.0 g drug, several times a day, equivalent preparations **ESCOP:** 2–3 g drug, 2–3 times a day, equivalent preparations

Stock plant (Latin)	Plant part	F	NF	MP	trad. MP	List A	List B	List C
Bixa orellana L., syn. Bixa POIR., Bixa platycarpa RUIZ et PAV.	Seed	× S						
Borago officinalis L.	Herb	× S					×	
Borago officinalis L.	Seed	×					×	
Boswellia serrata ROXB. ex COLEBR., syn. Boswellia glabra ROXB., Boswellia thurifera ROXB. ex FLEM., Libanus thuriferus COLEBR.	Resin		Not NFS	×			×	
Brassica nigra (L.) KOCH, syn. Brassica bracteoloata FISCH et MEY., Crucifera sinapis E.H.L. KRAUSE, Erysimum glabrum PRESL., Melanosinapis communis SCHIMP. et SPENN., Mutarda nigra BERNH., Sinapis nigra L.	Seed	× S						
Brassica oleracea L.	Inflorescence, stem	×						
Brassica oleracea L. var. suttoniana LEV., syn. Crucifera brassica E.H.L. KRAUSE	Leaf	×						
Brassica oleracea var. sabellica L.	Leaf	×						
Brugmansia spp.	all plant parts					×		
Bryonia spp.	Root			×		×		

Remarks	Risks	Critical plant substances	Pharmacologically effective dose
used as spice or colouring (Teuscher, Gewürz-drogen), extracts of it as additive E160b, colouring, restrictions of use as additive acc. to Annexes 1b, 1c of the ZZulV [Regulation on the Authorisation of Food Additives], as from 1 June 2012 use acc. to Reg. (EU) No. 1129/2011, specification acc. to Reg. (EU) No. 231/2012			
Negative monograph: due to risks, toxic II (Wink/Wyk), non-toxic to hardly toxic (+) (Roth/Daunderer), borage is a substance of concern as it contains pyrrolizidine (AMK), due to this pyrrolizidine content, it should only be used sparingly (Teuscher), BfArM graduated plan of 5 June 1992 concerning medicinal products containing pyrrolizidine	contains varying amounts of toxic pyrrolizidine alkaloids (PA), which are known to have an organotoxic, in particular hepatotoxic effect. PA demonstrated a carcinogenic effect with a genotoxic mechanism of action in animal experiments (Hager). All plant parts contain varying amounts of pyrrolizidine alkaloids.	Pyrrolizidine alkaloids	
toxic II (Wink/Wyk), licensed anthroposophic medicinal product, indications: haemorrhoids, varicose syndrome, borage is a substance of concern as it contains pyrrolizidine (AMK), BfArM graduated plan of 5 June 1992 concerning medicinal products containing pyrrolizidine, the cold-pressed oil of borage seeds is usually used as food	contains varying amounts of toxic pyrrolizidine alkaloids (PA), which are known to have an organotoxic, in particular hepatotoxic effect. PA demonstrated a carcinogenic effect with a genotoxic mechanism of action in animal experiments (Hager). All plant parts contain varying amounts of pyrrolizidine alkaloids. Cold-pressed borage oil contains no or only traces of pyrrolizidine alkaloids. (Frohne/Pfänder)	Pyrrolizidine alkaloids	
WHO monograph: orally for the management of arthritis, bronchial asthma, Crohn's disease and ulcerative colitis, **ESCOP monograph:** relief of painful osteoarthritis, symptomatic treatment of inflammatory bowel disease			**ESCOP:** 250–3,600 mg/day **WHO:** 1–3 g drug/day
Harmful in high doses (+) (Roth/Daunderer), toxic II–III (in high doses) (Wink/Wyk)	Ingestion of high doses causes gastrointestinal complaints, nausea, vomiting, diarrhoea, severe poisoning causes central paralysis, bradycardia and bradypnoea, even coma and death (Roth/Daunderer, Wink/Wyk)	Glucosinolates	
formerly belonged to the genus of Datura, extremely toxic +++ (Roth/Daunderer), extremely toxic Ia (Wink/Wyk)	psychotropic, CNS-inhibitor and sedative in low concentrations, ingestion of high doses causes hallucinations, euphoria, disorientation, insomnia and even death by respiratory arrest (Wink/Wyk)	Alkaloids (tropane alkaloids, e.g. L-hyoscyamine, L-scopolamine, atropine)	
Negative monograph: due to risks, use as drastic laxative and emetic obsolete, very toxic Ib (Wink/Wyk), highly toxic ++ (Roth/Daunderer)	Vertigo, vomiting, intense colic, severe watery, sometimes also bloody diarrhoea, gastroenteritis, kidney damage, abortion, cramps, death by respiratory arrest (Wink/Wyk, Roth/Daunderer)	Cucurbitacins	

Stock plant (Latin)	Plant part	F	NF	MP	trad. MP	List A	List B	List C
Buxus sempervirens L.	all plant parts					×		
Calendula officinalis L., syn. Calendula officinalis var. anemonaeflora HORT., Calendula officinalis var. hortensis FIORI, Calendula santamaria FONT-QUER, Caltha officinalis MOENCH	Flower	× T			×			
Calluna vulgaris (L.) HULL, syn. Calluna sagittaefolia GRAY, Erica vulgaris L.	Herb	× T						
Camellia sinensis (L.) O. KUNTZE, syn. Camellia assamica (J.W. MASTERS) W. WIGHT, Camellia chinensis (SIMS) KUNTZE, Thea bohea L., Thea viridis RAF.	Leaf, leaf bud	× T						
Cannabis sativa L.	Seed	×					×	
Capparis spinosa L.	Bud	× S						
Capsella bursa-pastoris (L.) MEDIK., syn. Bursa pastoris WIGG., Capsella polymorpha CAV., Iberis bursa-pastoris CRANTZ	Herb	× T		×	×		×	
Capsicum annuum L. convar. grossum	Fruit	×						
Capsicum frutescens L.	Fruit	× S						
Carica papaya L., syn. Carica hermaphrodita , Papaya carica GAERTN., Papaya vulgaris DC.	Leaf		Not NFS					×
Carica papaya L., syn. Carica hermaphrodita , Papaya carica GAERTN., Papaya vulgaris DC.	Fruit	×						
Carthamus tinctorius L.	Flower	× C		×			×	

Remarks	Risks	Critical plant substances	Pharmacologically effective dose
very toxic Ib (Wink/Wyk), highly toxic ++ (Roth/Daunderer)	Contact dermatitis, ingestion causes first excitement then paralysis and drop in blood pressure, nausea, vomiting, diarrhoea, vertigo, cramps circulatory collapse, death by respiratory arrest (Wink/Wyk)	steroidal alkaloids	
Positive monograph: for external use, **WHO monograph, HMPC monograph:** traditional use, **ESCOP monograph:** for external use only, **standard marketing authorisation:** 1209.99.99			**ESCOP:** external use only **WHO:** external use only **Commission E:** 1–2 g/150 ml for external, local use
Neutral monograph			
HMPC monograph: in progress (unfermented leaves)			
Cannabis sativa as well as the cannabionid Δ9-tetrahydrocannabinol (THC) are subject to the provisions of the BtMG [Narcotic Drugs Act]. According to Annex I to Section I Clause 1 BtMG, the seeds of C. sativa are exempted from this, unless they are intended for illegal cultivation. Processing in food production as well as distribution of products produced using Cannabis seeds is thus permitted under the Narcotic Drugs Act. Mildly toxic III (Wink/Wyk), toxic + (Roth/Daunderer)	euphorigenic and hallucinogenic effect, aphrodisiac, overdose causes cardiac complaints and CNS disorders, state of intoxication, illusion, mydriasis, impaired reactions, loss of sense of time, incorrect estimation of distance and speed, impaired concentration and attention, disorientation, impaired coordination, accelerated pulse, sudden blackout may occur after 1–3 months (Roth/Daunderer)	Cannabinoids (e.g. THC)	
	irritating, skin irritation may occur on contact with the plant (Roth/Daunderer)	Leek oils, glucosinolates	
Positive monograph, indications: mild menorrhagia, nasal bleeding, **HMPC monograph:** traditional use, **standard marketing authorisation:** 1539.99.99			**Commission E:** 10–15 g drug/day, equivalent preparations
HMPC monograph: in progress			
Positive monograph: for external use, **ESCOP monograph:** for external use, mildly toxic III (Wink/Wyk), slightly toxic + (Roth/Daunderer), **HMPC monograph:** in progress	chronic poisoning resulting from abuse (e.g. overdose as spice) is accompanied by lack of appetite, chronic gastritis, kidney and liver damage (Roth/Daunderer)	Capsaicin	
Neutral monograph			
WHO monograph: treatment of amenorrhoea, dysmenorrhoea and wounds or sores with pain and swelling, and prevention of atherosclerosis, used as colouring and colouring food (safflower)			**WHO:** 3–9 g drug/day

Stock plant (Latin)	Plant part	F	NF	MP	trad. MP	List A	List B	List C
Carum carvi L., syn. Apium carvi (L.) CRANTZ, Seseli carvi LAM., Ligusticum carvi ROTH.	Fruit	× T, S		×			×	
Cassia fistula L., syn. Bactyrilobium fistula WILLD., Cassia bonplandiana DC., Cassia excelsa H.B.K., Cassia fistuloides COLL., Cathartocarpus excelsus G. DON	Fruit	× F						
Cassia senna L. syn. Cassia acutifolia DEL., Senna acutifolia LINK and Cassia angustifolia VAHL. syn. Cassia lanceolata WIGHT et ARN., Senna alexandrina MILL., Senna angustifolia BATKA	Leaf, fruit			×		×		
Catha edulis FORSK., syn. Catha edulis (VAHL.) FORSK. ex ENDL, Celastrus edulis VAHL.	all plant parts			×		×		
Catharanthus roseus (L.) G. DON, syn.Vinca rosea L., Lochnera rosea (L.) RCHB., Ammocallis SMALL	all plant parts					×		
Caulophyllum thalictroides (L.) MICHX., syn. Leontice thalictroides L.	all plant parts					×		
Centaurea cyanus L., syn. Cyanus arvensis MOENCH, Cyanus segetum LAM.	Flower	× T						

Remarks	Risks	Critical plant substances	Pharmacologically effective dose
Positive monograph, indications: digestive problems such as mild gastrointestinal cramps, bloating and feeling of fullness, **standard marketing authorisation:** 1109.99.99, **ESCOP monograph:** spasmodic gastro-intestinal complaints, flatulence, bloating. Flatulent colic of infants, **HMPC monograph:** in progress			**Commission E:** 1.5–6 g drug/day, equivalent preparations **ESCOP:** 1.5–6 g drug/day
used to produce spirits			
Positive monograph, indications: short-term treatment of constipation, **standard marketing authorisation:** 7399.99.99 (senna leaves), 1259.99.99 (senna fruits, Alexandrian), 1269.99.99 (senna fruits, Tinnevelly), **WHO monograph (leaves and fruits):** short-term use in occasional constipation, **ESCOP monograph (leaves and fruits):** for short-term use in cases of occasional constipation, **HMPC monograph (leaves and fruits):** short-term treatment of occasional constipation; BfArM graduated plan of 21 June 1996 concerning medicinal products for human use containing anthranoid-containing plant genera, slightly toxic (+) (Roth/Daunderer), mildly toxic III (Wink/Wyk)	Electrolyte loss (in particular potassium) may be accompanied by muscular asthenia and cardiac dysfunction, albuminuria, haematuria, pseudomelanosis coli, aggravation of colonic inertia (development of laxative addiction) resulting from paralysis of intestinal muscles due to loss of potassium, contraindications and interactions reported (Commission E monograph), damage to intestinal nerves due to stimulation of uterine muscles, risk of abortion in pregnant women (Hager)	Anthranoids	**Commission E:** 20–30 mg hydroxyanthracene derivatives/day, calculated as sennoside B **ESCOP:** preparations equivalent to 15–30 mg hydroxyanthracene derivatives (calculated as sennoside B), no administration to children under the age of 10 **WHO:** 1–2 g drug/day equivalent to 10–30 mg sennosides, no administration to children under the age of 10 **HMPC:** equivalent to 15–30 mg hydroxyanthracene derivatives/day (calculated as sennoside B), no administration to children under the age of 12
Some plant substances are prescription-only and/or are subject to the provisions of the Narcotic Drugs Act (BtMG), mildly toxic II to toxic III (Wink/Wyk), toxic + (Roth/Daunderer)	psychotropic, effect similar to amphetamines, excitement, constipation, circulatory distress (Wink/Wyk, Roth/Daunderer)	Alkaloids (e.g. cathinone, cathine [=D-norpseudoephedrine])	
highly toxic ++ (Roth/Daunderer), toxic Ib to very toxic II (Wink/Wyk), BfArM graduated plan of 20 July 1987 concerning medicinal products containing Madagascar rosy periwinkle	mitotic inhibition, neurological disorders, nausea, vomiting, paraesthesia of hands and feet, visual disturbances and hearing, fever, inhibition of leucocytopoiesis, loss of hair, impairment of intestinal mucosa, constipation, diarrhoea, neuromuscular vegetative disorders, affects CNS and mental state, apoplexy (Wink/Wyk, Roth/Daunderer)	Indole alkaloids (Vinca alkaloids)	
toxic Ib to very toxic II (Wink/Wyk), toxic + (Roth/Daunderer)	causes diarrhoea, vomiting and gastroenteritis in low doses, ingestion of high doses causes burning in the mouth, hypersalivation, sweating, mydriasis, uterus contractions, nausea, paralysis, also hallucinations, delirium, respiratory arrest (Wink/Wyk), risk of abortion during pregnancy reported, oestrogenic and antispasmodic effects reported (Roth/Daunderer)	Quinolizidine alkaloids, saponins	
Neutral monograph			

Stock plant (Latin)	Plant part	F	NF	MP	trad. MP	List A	List B	List C
Centaurium erythraea RAFN, syn. Centaurium minus GARSAULT, Centaurium minus MOENCH P.P., Centaurium umbellatum GILIBERT, Erythraea centaurium BORKH.	Herb	× T		×	×		×	
Cephaelis asiatica (L.) URBAN, syn. Centella asiatica (L.) NANNF., Centella triflora (R. et P.) NANNF., Hydrocotyle asiatica L.	Herb	×		×			×	
Cephaelis ipecacuanha (BROT.) A. RICH., syn. Uragoga ipecacuanha (WILLD.) BAILL., Psychotria ipecacuanha (BROT.) MUELL. ARG. (auch STOKES)	Root			×		×		
Ceratonia siliqua L., syn. Ceratonia coriacea SALISB., Ceratonia inermis STOKES	Fruit	×						
Chamaemelum nobile (L.) ALL., syn. Anthemis nobilis L.	Flower	× T			×			
Chelidonium majus L., syn. Chelidonium grandiflorum DC., Chelidonium japonicum THUNB., Chelidonium luteum GILIB., Chelidonium umbelliferum STOCK.	Herb			×		×		
Chenopodium ambrosioides L. var. Anthelminticum A. GRAY, syn. Chenopodium ambrosioides L. ssp. eu-ambrosioides AELLEN var. anthelminticum AELLEN	all plant parts			×		×		
Chondrodendron tomentosum RUIZ et PAV., syn. Botryopsis platyphylla BENTH.	all plant parts			×		×		
Cicer arietinum L.	Seed	×						

Remarks	Risks	Critical plant substances	Pharmacologically effective dose
Positive monograph, indications: lack of appetite, digestive problems **standard marketing authorisation:** 1319.99.99, **ESCOP monograph:** dyspeptic complaints, lack of appetite, **HMPC monograph:** traditional use			**Commission E:** average daily dose: 6 g drug, equivalent preparations **ESCOP:** 1–4 g drug/150 ml water, up to 3 times a day
WHO monograph, indications: extracts have been administered orally to treat stress-induced stomach and duodenal ulcers, toxic II (Wink/Wyk), toxic + (Roth/Daunderer), **ESCOP monograph, indications:** chron. venous insufficiency, varicosis, wound healing, **HMPC assessment report**	Irritation of skin and mucous membranes, ingestion of higher doses may cause digestive problems (Wink/Wyk)	Triterpene saponins, hydrocotyline (alkaloid)	**WHO:** 0.33–0.68 g drug, 3 times a day, equivalent preparations **ESCOP:** 0.6 g drug, up to 4 times a day
WHO monograph, indications: a syrup made from the roots is used as an emetic, to empty the stomach in cases of poison ingestion, extremely toxic +++ (Roth/Daunderer), emetines prescription-only, Cephaelis ipecacuana and Cephaelis acuminata pharmacy-only, used in medicinal products as an ingredient of expectorants and emetics	Pneumonia, nausea, vomiting, bloody diarrhoea, pain, exhaustion, bradycardia, myoplegia, anuria (Roth/Daunderer)	Alkaloids (e.g. cephaeline, emetine)	**WHO:** 21–42 mg total alkaloid content (special doses for children)
Carob bean gum additive E 410			
Negative monograph: due to risks, **HMPC monograph:** traditional use	Allergic reactions may occur (Commission E monograph)		
Positive monograph, indications: cramps in the biliary and gastrointestinal tract, **WHO monograph:** used for the symptomatic treatment of mild to moderate spasms of the upper gastrointestinal tract, minor gallbladder disorders, and dyspeptic complaints such as bloating and flatulence, **ESCOP monograph:** symptomatic treatment of mild to moderate spasms of the upper gastrointestinal tract, minor gall bladder disorders, dyspeptic complaints such as bloating and flatulence; BfArM graduated plan of 9 Apr 2008 concerning medicinal products for internal use containing celandine, http://www.bfarm.de/DE/Pharmakovigilanz/risikoinfo/stufenplanverf/Liste/stp-schoellkraut.html, highly toxic ++ (Roth/Daunderer), toxic II (Wink/Wyk), **HMPC assessment report**	Inflammations and burning in the mouth, vomiting, paralysis, vesical tenesmus, dizziness, arrhythmia, gastrointestinal complaints accompanied by bloody diarrhoea, bradycardia, drop in blood pressure, finally death by circulatory collapse, hepatotoxic (Wink/Wyk, Roth/Daunderer)	Isoquinoline alkaloids (e.g. chelidonine, sanguinarine, berberine)	**Commission E:** 2–5 g drug/day **WHO:** 2–5 g drug/day **ESCOP:** 1.2–3.6 g/day
toxic + (plant), highly toxic ++ (oil) (Roth/Daunderer), very toxic, Ib (Wink/Wyk), information: oil prescription-only (Chenopodii anthelminithici oleum)	Ingestion of higher doses causes CNS disorders, loss of consciousness, paralysis, drop in blood pressure, increased tendency to haemorrhage, cerebral haemorrhage, damage to the cochlear nerve, impaired hearing, digestive tract inflammations accompanied by cramps, abortifacient (Roth/Daunderer)	Terpenes (monoterpenes e.g. ascaridole)	
extremely toxic Ia (Wink/Wyk), pareira and preparations thereof prescription-only	strong muscle relaxant, hypotensive, circulatory collapse, bronchospasm, death by cardiac arrest or respiratory arrest (Wink/Wyk)	Alkaloids (e.g. tubocurarine)	

Stock plant (Latin)	Plant part	F	NF	MP	trad. MP	List A	List B	List C
Cichorium intybus L. var. foliosum HEGI	Leaf bud	×						
Cichorium intybus L. var. intybus = ssp. intybus var. sylvestre VISIANI = C. sylvestre G. BANK	Herb, root	× T		×	×		×	
Cichorium intybus L. var. sativum LAM. et DC. = ssp. sativum (DC.) JANCHEN var. radicosum ALEF.	Root	×						
Cicuta virosa L., syn. Cicuta angustifolia KIT., Cicuta aquatica DUMORT., Selinum virosum (L.) E.H. KRAUSE	all plant parts			×		×		
Cimicifuga racemosa (L.) NUTT., syn. Actaea racemosa L., Botrophis actaeoides RAFIN.	Rhizome			×		×		
Cinchona pubescens VAHL, Cinchona succirubra PAVON ex KLOTSCH	Bark	× F		×			×	
Cinnamomum cassia NEES ex BL., syn. Cinnamomum aromaticum NEES, Laurus cassia NEES	Flower	× S						
Cinnamomum cassia NEES ex BL., syn. Cinnamomum aromaticum NEES, Laurus cassia NEES	Bark	× S		×			×	
Cinnamomum ceylanicum BL., syn. Cinnamomum verum J.S. PRESL, Laurus cassia BURM.	Bark	× S		×	×		×	

Remarks	Risks	Critical plant substances	Pharmacologically effective dose
Positive monograph, indications: lack of appetite, digestive problems, **HMPC monograph:** traditional use			**Commission E:** 3 g drug/day, equivalent preparations
Pharmacy-only (only homeopathic drugs on the market), very toxic Ib (Wink/Wyk), extremely toxic +++ (Roth/Daunderer)	Burning in mouth and throat, nausea, vomiting, gastric pain, mydriasis, headache, tremor, bradycardia, epileptic fits, delirium, coma, paralysis, death by respiratory arrest (Wink/Wyk, Roth/Daunderer)	Polyacetylenes (e.g. cicutoxin, cicutol)	
Positive monograph, indications: premenstrual and dysmenorrhoeal as well as climacteric neurovegetative symptoms, **WHO monograph:** treatment of climacteric symptoms such as hot flushes, profuse sweating, sleeping disorders and nervous irritability, **ESCOP monograph:** climacteric symptoms such as hot flushes, profuse sweating, sleep disorders and nervous irritability, BfArM graduated plan of 9 June 2009 concerning medicinal products containing Cimicifuga, **HMPC monograph:** herbal medicinal product for the relief of menopausal complaints such as hot flushes and profuse sweating	hepatotoxic, s. HMPC statement: http://www.bfarm.de/SharedDocs/1_ Downloads/DE/Pharmakovigilanz/ stufenplverf/cimicifuga_hmpc_ assessment.pdf?__blob=publicationFile, s. BfArM notification: http:// www.bfarm.de/SharedDocs/1_ Downloads/DE/Pharmakovigilanz/ stufenplverf/cimicifuga_bescheid. pdf?__blob=publicationFile	Triterpene glycosides, isoquinoline alkaloids (e.g. cytisine)	**Commission E, WHO:** 40–60 % alcoholic extract, equivalent to 40 mg drug
Positive monograph, indications: lack of appetite, digestive problems such as feeling of fullness and bloating, **standard marketing authorisation:** 1459.99.99, toxic II (Wink/ Wyk), toxic + (Roth/Daunderer), source material for quinine, dose restriction for quinine acc. to the Flavouring Regulation, BfR statement on quinine: http://www.bfr.bund.de/ cm/208/chininhaltige_getraenke_koennen_ gesundheitliich_problematisch_sein.pdf	Overdose may cause fever, icterus, haematuria, vomiting, diarrhoea, shivering, dyspnoea, internal haemorrhage and respiratory arrest (Wink/Wyk, Roth/Daunderer)	Alkaloids (e.g. quinine, quinidine)	**Commission E:** 1–3 g drug/day
Negative monograph: due to risks, use as taste improver is considered to be safe (Commission E monograph)	frequently induces allergic reactions of skin and mucous membranes, contraindications reported (Commission E monograph)	Cinnamic aldehyde	
Positive monograph, indications: lack of appetite, digestive problems such as mild gastrointestinal cramps, feeling of fullness or bloating, s. BfR statement: http://www.bfr. bund.de/cm/343/neue-erkenntnisse-zu-cumarin-in-zimt.pdf		Cinnamic aldehyde, coumarin	**Commission E, WHO:** 2–4 g drug/day, equivalent preparations
Positive monograph, indications: lack of appetite, digestive problems such as mild gastrointestinal cramps, feeling of fullness or bloating, **ESCOP monograph:** dyspeptic complaints such as gastrointestinal spasms, bloating and flatulence, loss of appetite, diarrhoea, **WHO monograph, HMPC monograph:** traditional use, **standard marketing authorisation:** 1709.99.99, s. BfR statement: http://www.bfr.bund.de/cm/343/neue-erkenntnisse-zu-cumarin-in-zimt.pdf		Cinnamic aldehyde, coumarin	**Commission E, WHO:** 2–4 g drug/day, equivalent preparations

Stock plant (Latin)	Plant part	F	NF	MP	trad. MP	List A	List B	List C
Cistus x incanus L.	all plant parts		Not NFS	×			×	
Citrullus colocynthis (L.) SCHRAD., syn. Cucumis colocynthis L., Colocynthis officinalis SCHRAD.	all plant parts			×		×		
Citrus aurantifolia (CHRISTM.) SWINGLE and Citrus latifolia TANAKA	Peel	× T						
Citrus aurantifolia (CHRISTM.) SWINGLE, syn. Citrus aurantiifolia (CHRISTM. et PANZ.) SWINGLE, Limonia acidissima HOUTT. (non L.), Limonia aurantifolia CHRISTM.	Fruit	×						
Citrus aurantium L. ssp. aurantium, syn. Citrus aurantium L. ssp. amara (L.) ENGL, Citrus communis LEMAOUT. & DECAISN., Citrus florida SALISB.	Leaf	× T						
Citrus aurantium L. ssp. aurantium, syn. Citrus aurantium L. ssp. amara (L.) ENGL, Citrus communis LEMAOUT. & DECAISN., Citrus florida SALISB.	Flower	× C						
Citrus aurantium L. ssp. aurantium, syn. Citrus aurantium L. ssp. amara (L.) ENGL, Citrus communis LEMAOUT. & DECAISN., Citrus florida SALISB.	Fruit	×						
Citrus aurantium L. ssp. aurantium, syn. Citrus aurantium L. ssp. amara (L.) ENGL, Citrus communis LEMAOUT. & DECAISN., Citrus florida SALISB.	Peel	×		×			×	
Citrus latifolia TANAKA	Fruit	×						
Citrus limon (L.) BURM.f., syn. Citrus limonum RISSO, Limon vulgaris MILL.	Fruit	×						
Citrus limon (L.) BURM.f., syn. Citrus limonum RISSO, Limon vulgaris MILL.	Peel	× S, T						
Citrus paradisi MACF., syn. Citrus decumana var. patoniaca RICCOB., Citrus decumana var. racemosa ROEM.	Fruit	×						

Remarks	Risks	Critical plant substances	Pharmacologically effective dose
Notification of the Federal Institute for Drugs and Medical Devices acc. to Section 21 Clause 4 AMG [Medicinal Products Act] of 4 Mar 2008 and decision of the North-Rhine Westphalia Higher Administrative Court of 15 Mar 2010 (file ref: 13 A 2612/09) on the classification of products containing preparations of Cistus incanus as medicinal products			
Negative monograph: due to risks, highly toxic ++ (Roth/Daunderer), very toxic Ib (Wink/Wyk)	cytotoxic, irritating to gastrointestinal mucosa up to bloody diarrhoea, kidney damage, haemorrhagic cystitis, cramps, abortifacient effect (Wink/ Wyk, Roth/Daunderer)	Cucurbitacins	
	concerning synephrine s. BfR statement: http://www.bfr.bund. de/cm/343/gesundheitliche-bewertung-von-synephrin-und-koffeinhaltigen-sportlerprodukten-und-schlankheitsmitteln.pdf	Synephrine	
Positive monograph, indications: lack of appetite, digestive problems, **standard marketing authorisation:** 1629.99.99, hardly toxic (Roth/Daunderer)	phototoxic, concerning synephrine s. BfR statement: http://www.bfr. bund.de/cm/343/gesundheitliche-bewertung-von-synephrin-und-koffeinhaltigen-sportlerprodukten-und-schlankheitsmitteln.pdf	Furocoumarins	**Commission E:** 4–6 g drug/day, equivalent preparations

Stock plant (Latin)	Plant part	F	NF	MP	trad. MP	List A	List B	List C
Citrus paradisi MACF., syn. Citrus decumana var. patoniaca RICCOB., Citrus decumana var. racemosa ROEM.	Peel	× F						
Citrus reticulata BLANCO, syn. Citrus deliciosa TEN., Citrus nobilis ANDR. (non LOUR.), Citrus nobilis var. genuina TANAKA, Citrus nobilis var. major KERR.	Fruit	×						
Citrus reticulata BLANCO, syn. Citrus deliciosa TEN., Citrus nobilis ANDR. (non LOUR.), Citrus nobilis var. genuina TANAKA, Citrus nobilis var. major KERR.	Peel	× T						
Citrus sinensis (L.) OSBECK, syn. Aurantium sinensis MILL., Citrus aurantium LOUR. (non L.)	Leaf	× T						
Citrus sinensis (L.) OSBECK, syn. Aurantium sinensis MILL., Citrus aurantium LOUR. (non L.)	Flower	× C						
Citrus sinensis (L.) OSBECK, syn. Aurantium sinensis MILL., Citrus aurantium LOUR. (non L.)	Fruit	×						
Citrus sinensis (L.) OSBECK, syn. Aurantium sinensis MILL., Citrus aurantium LOUR. (non L.)	Peel	× T, S		×			×	
Cnicus benedictus L., syn. Calcitrapa benedicta LAM., Carbenia benedicta BENTH. et HOOK.	Herb	× T		×	×		×	
Cocos nucifera L., syn. Calappa nucifera KUNTZE, Cocos mamillaris BLANCO	Seed	×						
Coffea arabica L., syn. Coffea laurifolia SALISB., Jasminum arabicum laurifolia DEJUSS.	Seed	×						
Cola nitida (VENT.) SCHOTT et ENDL., syn. Cola acuminata ENDL.	Seed	×		×	×		×	
Colchicum autumnale L., syn. Colchicum commune NECK., Colchicum crociflorum ANDERS	all plant parts			×		×		
Colutea arborescens L., syn. Colutea vesicaria SEGUIER	all plant parts					×		

Remarks	Risks	Critical plant substances	Pharmacologically effective dose
Positive monograph, indications: lack of appetite			**Commission E:** 10–15 g drug/day, equivalent preparations
Positive monograph, indications: lack of appetite, digestive problems, hardly toxic but sensitising (Roth/Daunderer)	Cnicin contained in the drug causes heat and burning in the throat and the oesophagus, nausea, colic and diarrhoea in low doses (Roth/Daunderer), allergic reactions may occur (Commission E monograph)	Sesquiterpene lactones (e.g. cnicin)	**Commission E:** 4–6 g drug/day, equivalent preparations
Positive monograph, indications: mental and physical fatigue, **HMPC monograph:** traditional use, mildly toxic III (Wink/Wyk), slightly toxic (Roth/Daunderer)	Ingestion of higher doses causes headache, vertigo, sleep disturbances, agitation and tachycardia (Wink/Wyk)	Purine alkaloids (e.g. caffeine, theobromine)	**Commission E:** 2–6 g drug/day
Positive monograph, indications: acute attacks of gout, familial Mediterranean fever, cytotoxic, extremely toxic Ia (Wink/van Wyk), extremely toxic +++ (Roth/Daunderer)	Nausea, vomiting, watery and bloody diarrhoea, intense abdominal pain, haematuria, drop in blood pressure, convulsions, paralysis, overdose causes death by respiratory and circulatory arrest (Wink/Wyk), contraindications and interactions reported (Commission E monograph)	Colchicum alkaloids (e.g. colchicine)	**Commission E:** 0.5–1.5 mg colchicine/day, max. dose = 8 mg colchicine
toxic II (Wink/Wyk), toxic + (Roth/Daunderer)	Diarrhoea, sometimes vomiting (Wink/Wyk)	L-canavanine, lectins	

Stock plant (Latin)	Plant part	F	NF	MP	trad. MP	List A	List B	List C
Conium spp.	all plant parts			×		×		
Consolida regalis S.F. GRAY, syn. Delphinium consolida L.	Flower	× C						
Convallaria majalis L., syn. Convallaria fragrans SALISB., Convallaria latifolia MILL., Polygonatum majale ALL.	Herb			×		×		
Convolvulus scammonia L., syn. Convolvulus syriacus MORIS, Convolvulus pseudoscammonia C. KOCH	all plant parts			×		×		
Coriandrum sativum L., syn. Coriandrum maius GOUAN, Coriandrum diversifolium GILIB.	Herb	× S, T						
Coriandrum sativum L., syn. Coriandrum maius GOUAN, Coriandrum diversifolium GILIB.	Fruit	× S		×			×	
Corylus avellana L., syn. Corylus silverstris SALISB.	Leaf	× T						
Corylus avellana L., syn. Corylus silverstris SALISB.	Seed	×						
Crataegus laevigata (POIRET) D.C., syn. Crataegus oxyacantha L.P.P. et AUCT.	Leaf, flower	× T		×			×	
Crataegus laevigata (POIRET) D.C., syn. Crataegus oxyacantha L.P.P. et AUCT.	Fruit	×		×			×	

Remarks	Risks	Critical plant substances	Pharmacologically effective dose
extremely toxic +++ (Roth/Daunderer), extremely toxic Ia (Wink/Wyk), used only in homeopathic drugs	all plant parts extremely toxic, paralysis of motor nerve ends involving the following symptoms: burning in mouth and throat, paralysis of tongue, vomiting, hypersalivation, diarrhoea, tachycardia, dyspnoea, disorientation, ascending paralysis, cold spills and numbness, death by apnoea usually while the patient is fully conscious (Wink/Wyk, Roth/Daunderer)	Piperidine alkaloids (e.g. coniine)	
Neutral monograph (for use as colouring in amounts of up to 1 %): poisonous plant, toxic + (Roth/Daunderer), extremely toxic Ia (Wink/Wyk), the (toxic) alkaloids are found in the underground parts, the tops and the seeds (Rimpler). Reliable information about the alkaloid content in the flowers is not available; according to the literature, their use as colouring in tea blends, i.e. in amounts of <1 %, is considered to be safe (BGA monograph)	Nausea, excitement, cardiac arrhythmia, cramps, gastroenteritis, respiratory arrest (Roth/Daunderer, Wink/Wyk), relevant diterpene alkaloids are found especially in seeds and roots (Hager)	Alkaloids (e.g. delphinine)	
Positive monograph, indications: mild exercise insufficiency, senile heart disease, chronic pulmonary heart disease, lily-of-the-valley plant itself pharmacy-only, extremely toxic +++ (Roth/Daunderer), very toxic Ib (Wink/Wyk)	Nausea, vomiting, diarrhoea, arrhythmia, hypertension, coma, cardiac arrest, interactions with medicinal products (Wink/Wyk, Roth/Daunderer, Schilcher)	cardioactive glycosides (e.g. convallatoxin)	**Commission E:** 0.6 g/day standardised lily-of-the-valley powder, equivalent preparations
drastic laxative		Glycoretins	
Positive monograph, indications: digestive problems, lack of appetite, **standard marketing authorisation:** 1079.99.99			**Commission E:** 3 g drug/day, equivalent preparations
Positive monograph for combination only (Crataegi folium cum flore), **indications:** decreasing cardiac output equivalent to stage II acc. to NYHA, active ingredients: flavonoids (e.g. hyperoside, rutin and OPC), effect demonstrated for standardised extracts from leaves with flowers, **ESCOP monograph and WHO monograph for combinations (Crataegi folium cum flore):** declining cardiac performance corresponding to Functional Capacity Class II as defined by the NYHA, nervous heart complaints, support of cardiac and circulatory functions, **HMPC monograph:** in progress			**Commission E:** 160–900 mg native, water-alcohol extract (ethanol 45 % or methanol 70 %; DER 4–7:1 with defined flavonoid or procyanidin content) equivalent to 30–168.7 mg oligomer procyanidins calculated as epicatechin or 3.5–19.8 mg flavonoids calculated as hyperoside in 2 or 3 single doses **ESCOP, WHO:** 1–1.5 g drug, 3–4 times a day
Neutral monograph, ESCOP monograph: cardiac complaints, support of cardiac and circulatory functions, **standard marketing authorisation:** 1349.99.99, **HMPC monograph:** in progress			**ESCOP:** 0.3–1 g drug, 3 times a day

Stock plant (Latin)	Plant part	F	NF	MP	trad. MP	List A	List B	List C
Crocus sativus L., syn. Crocus autumnalis SM., Crocus officinalis MARTYN.	Stigma, style	× S		×			×	
Croton spp.	all plant parts			×		×		
Cucumis melo L., syn. Cucumis acidus JACQ., Cucumis deliciosus SALISB., Melo agrestis PANG., var. reticulata and var. cantalupa and var. inodorus	Fruit	×						
Cucurbita pepo L., syn. Cucurbita courgero SER., Pepo melopepo MOENCH	Seed	×		×	×		×	
Cuminum cyminum L., syn. Cuminia cyminum J.F. GMELIN	Fruit	× S						
Curcuma longa L., syn.Curcuma domestica VAL.	Rhizome	× S		×	×		×	

Remarks	Risks	Critical plant substances	Pharmacologically effective dose
Negative monograph: due to risks, **WHO monograph:** tonic, sedative, emmenagogue, prevention of arteriosclerosis, also used as food, psychotropic, toxic II (Wink/Wyk), toxic + (Roth/Daunderer)	As much as 5 g of stigma may cause severe symptoms. Ingestion of as much as 10 g per day causes abortion and uterine haemorrhage, whereas a dose of 20 g is lethal. In low doses typically used in cuisine, saffron is harmless. Ingestion of higher doses causes uterine haemorrhage, bloody diarrhoea, haematuria and bleeding eyelids, effects on CNS such as excitement accompanied by fits of laughter, followed by vomiting, delirium, visual disturbances, lethargy, tachycardia, cardiac palpitation, central paralysis and death. Skin and mucous membranes take on a yellowish colour (Wink/Wyk)	Pinenes, eucalyptol (=1,8-cineol), safranal	**WHO:** 3–9 g drug/day, doses below 5 g/day recommended due to toxicity
extremely toxic +++ (Roth/Daunderer), extremely toxic Ia (Wink/Wyk), use of seed oil as a medicinal product, e.g. as a drastic laxative, obsolete, Croton cascarilla, Croton eleuteria and Croton tiglium pharmacy-only, substance of concern (AMK)	very toxic seeds, drastic laxative effect due to their croton oil content. As many as 4 seeds or approx. 20 drops of oil are fatal to humans, burning and tickle in the mouth, vomiting, oppression and agitation, watery, sometimes bloody diarrhoea, dizziness, vertigo, circulatory collapse, arrhythmia, nephritis, delirium, cocarcinogenic effect (Wink/Wyk, Roth/Daunderer)	Phorbol ester, tiglic acid	
Positive monograph, indications: irritable bladder, micturition difficulties resulting from stage I–II prostatic adenoma, active ingredients: amongst others cucurbitin, phytosterol, **WHO monograph:** for symptomatic treatment of difficulties with micturition associated with stage I–II prostatic adenoma and irritable bladder, **ESCOP monograph:** symptomatic treatment of micturition disorders (such as dysuria, pollakisuria, nocturia, urine retention and hesitancy) in benign prostatic hyperplasia at stages I and II as defined by Alken or stage II and III as defined by Vahlensieck, irritable bladder, **HMPC monograph:** traditional use			**Commission E, WHO:** 10 g drug/day, equivalent preparations **ESCOP:** 10–20 g drug/day
Positive monograph, indications: digestive problems, **WHO monograph:** the principal use of Rhizoma Curcumae Longae is for the treatment of acid, flatulent, or atonic dyspepsia, **ESCOP monograph:** symptomatic treatment of mild digestive disturbances and minor biliary dysfunction, **HMPC monograph:** traditional use, **standard marketing authorisation:** 2339.99.99			**Commission E, ESCOP, WHO:** 1.5–3 g drug/day, equivalent preparations

Stock plant (Latin)	Plant part	F	NF	MP	trad. MP	List A	List B	List C
Curcuma xanthorrhiza ROXB., syn. Curcuma zanthorrhiza ROXB.	Root	× S		×	×		×	
Curcuma zedoaria (BERG) ROSC., syn. Amomum latifolium LAM., Amomum zedoaria WILLD., Amomum zerumbeth KOEN. non L., Curcuma speciosa HORT. ex LINK., Curcuma zerumbeth (KOEN.) ROXB.	Rhizome	× S						
Cyclamen spp.	Root tuber					×		
Cyclopia genistoides (THUNB.) R. BR.	Herb	× T						
Cydonia oblonga MILLER, syn. Cydonia communis POIRET ex DUHAMEL, Cydonia vulgaris DE-LARBRE, Pyrus cydonia L., Sorbus cydonia CRANTZ	Fruit	×						
Cymbopogon citratus (DC.) STAPF, syn. Andropogon citratus DC., Andropogon ceriferus HACK., Andropogon citriodorum DESF., Andropogon roxburghii NEES ex STEUD.	Herb	× S, T						
Cynara scolymus L., syn. Cynara cardunculus L. ssp. Scolymus (L.) HAY	Foliage leaf		Not NFS	×	×		×	
Cynara scolymus L., syn. Cynara cardunculus L. ssp. Scolymus (L.) HAY	Receptacle, scale leaf	×			× (buds)			
Cynodon dactylon (L.) PERS., syn. Capriola dactylon (L.) KTZE., Panicum dactylon L., Digitaria stolonifera SCHRAD., Dactylon officinale VILL., Tibichia umbellata KOELER	Rhizome		Not NFS					×
Cynoglossum officinale L., syn. Cynoglossum clandestinum DESF.	Herb			×		×		

Remarks	Risks	Critical plant substances	Pharmacologically effective dose
Positive monograph, indications: digestive problems, **ESCOP monograph:** symptomatic treatment of mild digestive disturbances and minor biliary dysfunction, **HMPC monograph:** in progress	Long-term use causes gastric complaints, contraindications reported (Commission E monograph)	Curcuminoids, xanthorrhizol	**Commission E, ESCOP:** 2 g drug/day, equivalent preparations
Neutral monograph			
toxic II (Wink/Wyk), toxic + (Roth/Daunderer)	Nausea, vomiting, gastric pain, diarrhoea, sweats, circulatory distress, cramps, symptoms of paralysis, haemolysis Local irritations also reported (Roth/Daunderer)	Saponins (e.g. cyclamine)	
Neutral monograph			
Positive monograph, indications: digestive problems, **WHO monograph:** treatment of digestive complaints (e.g. dyspepsia, feeling of fullness, flatulence, nausea, stomach ache and vomiting). Adjunct treatment of mild to moderate hypercholesterolaemia, **ESCOP monograph:** digestive complaints (e.g. stomach ache, nausea, vomiting, feeling of fullness, flatulence) and hepatobiliary disturbance, adjuvant to a low fat diet in the treatment of mild to moderate hyperlipidaemia, **HMPC monograph:** traditional use			**Commission E:** 6 g drug/day, equivalent preparations **ESCOP:** 5–10 g dried drug/day (no administration to children) **WHO:** 1–10 g drug/day, equivalent preparations, not recommended to children under the age of 12
Negative monograph: due to risks, substance of concern (AMK), BfArM graduated plan of 5 June 1992 concerning medicinal products containing pyrrolizidine alkaloids	contains pyrrolizidine alkaloids with hepatotoxic, carcinogenic, partly also mutagenic and teratogenic properties (Frohne/Pfänder)	Pyrrolizidine alkaloids	

Stock plant (Latin)	Plant part	F	NF	MP	trad. MP	List A	List B	List C
Cytisus scoparius (L.) LINK, syn. Genista angulata POIRET, Sarothamnus scoparius (L.) WIMMER ex. KOCH, Spartium scoparium L.	all plant parts			×		×		
Daphne spp.	all plant parts					×		
Datura spp.	all plant parts			×		×		
Daucus carota L. ssp. sativus (HOFFM.) ARCANG.	Root	×						
Delphinium elatum L., syn. Delphinium intermedium AIT.	all plant parts					×		
Delphinium staphisagria L., syn. Delphinium officinale WENDER., Staphydis agria	all plant parts			×		×		
Dendranthema grandiflora TZVELEV, syn. Chrysanthemum morifolium RAMAT., Dendranthema grandiflora (RAMAT.) KITAM., Dendranthema morifolia (RAMAT.) TZVELEV	Flower		×					
Dendrocalamus asper (SCHULTES f.) HEYNE and Dendrocalamus latiflorus MUNRO, syn, Bambusa latiflora (MUNRO) KURZ	Sprout	× b						
Digitalis spp.	all plant parts			×		×		

Remarks	Risks	Critical plant substances	Pharmacologically effective dose
Negative monograph: due to risks associated with Scotch broom flowers, **Positive monograph:** for Scotch broom, indications: functional cardiovascular disorders, **standard marketing authorisation:** 1439.99.99, toxic II (Wink/Wyk), toxic + (all plant parts), (Roth/Daunderer)	circulatory collapse, arrhythmia, tachycardia, sweating, vomiting, diarrhoea, dizziness, vertigo, convulsions, headache, ascending paralysis, death by apnoea, hypertonia in pregnancy, contraindications reported (Wink/Wyk)	Alkaloids (e.g. sparteine)	**Commission E:** 1–1.5 g drug/day, water-ethanol extract
poisonous plant, extremely toxic +++ (Roth/Daunderer), extremely toxic Ia (Wink/Wyk)	Burning, swelling, blistering and redness in the mouth, hypersalivation, swallowing difficulties, nausea, bloody vomiting and bloody diarrhoea, disturbance of consciousness, fever, cramps, paralysis, kidney damage, tachycardia, circulatory collapse and death. After skin contact, there is a risk of swelling and blistering preceded by itching and redness (Wink/Wyk, Roth/Daunderer)	Phorbol ester (e.g. mezerein), cumaroyl glycoside (e.g. daphnin)	
Negative monograph: due to risks associated with the leaves and seeds of Datura stramonium, isolated alkaloids prescription-only, Datura species pharmacy-only, extremely toxic +++ (Roth/Daunderer), extremely toxic Ia (Wink/Wyk)	psychotropic, calmative and sedative in low doses, ingestion of higher doses causes intense hallucinations, excitement, facial blush, dry mouth, euphoria, mydriasis, disorientation, insomnia, respiratory arrest and death (Wink/Wyk)	Tropane alkaloids (e.g. atropine, L-hyoscyamine, scopolamine)	
	sometimes irritating, phototoxic	Furocoumarins	
poisonous plant, extremely toxic +++ (Roth/Daunderer), extremely toxic Ia (Wink/Wyk)	Nausea, excitement, cardiac arrhythmia, cramps, gastrointestinal tract inflammations, muscular asthenia, cardiac arrest, central paralysis, dyspnoea, death by myoplegia and cardioplegia (Wink/Wyk)	Terpene alkaloids (e.g. delphinine)	
Negative monograph: due to risks, extremely toxic +++ (Roth/Daunderer), extremely toxic Ia (Wink/Wyk) Delphinium staphisagria (stavesacre) pharmacy-only	Nausea, excitement, cardiac arrhythmia, cramps, gastroenteritis, respiratory arrest (Roth/Daunderer, Wink/Wyk), relevant diterpene alkaloids are found especially in seeds and roots (Hager)	Terpene alkaloids (e.g. delphinine, staphisine)	
toxic II (Wink/Wyk), cook before consumption	When injured, the sprouts release hydrocyanic acid (HCN). HCN inhibits cellular respiration in mitochondria. Symptoms of HCN poisoning. HCN must be eliminated before consumption (by cooking) (Wink/Wyk)	cyanogenic glycosides	
extremely toxic Ia (Wink/Wyk), extremely toxic +++ (Roth/Daunderer)	typical cardiac glycoside poisoning, very narrow therapeutic range, bradycardia, cardiac arrest, cardiac arrhythmia, nausea accompanied by constant retching, diarrhoea, gastroenteritis, visual disturbances, agitation, headache, convulsions, CNS disorders, disorientation, delirium, hallucinations, drop in blood pressure, disturbance of consciousness, sudden death (Wink/Wyk, Roth/Daunderer)	cardioactive glycosides (e.g. digoxin, digitoxin)	

Stock plant (Latin)	Plant part	F	NF	MP	trad. MP	List A	List B	List C
Dipteryx odorata (AUBL.) WILLD., syn. Coumarouna odorata AUBL., Baryosma tongo GAERTN.	Seed	× F						
Dracocephalum moldavica L.	Herb	× S						
Drosera spp.	Herb	× T		×			×	
Dryopteris filix-mas (L.) SCHOTT, syn. Aspidium filix-mas (L.) SW., Lastrea filix-mas (L.) PRESL., Polypodium filix-mas L.	all plant parts			×		×		
Ecballium elaterium (L.) A. RICH., syn. Ecballium officinale NEES, Momordica elaterium L.	all plant parts					×		
Echinacea angustifolia DC., syn. Brauneria angustifolia (DC.) A. HELLER, Echinacea pallida var. angustifolia (DC.) CRONQ.	Herb	× T						
Echinacea angustifolia DC., syn. Brauneria angustifolia (DC.) A. HELLER, Echinacea pallida var. angustifolia (DC.) CRONQ.	Root	× T		×	×		×	
Echinacea pallida (NUTT.) NUTT., syn. Brauneria pallida BRITTON, Echinacea angustifolia HOOKER, Echinacea pallida (NUTT.) NUTT. f. albida STEYERM., Rudbeckia pallida NUTT.	Herb		Not NFS					×
Echinacea pallida (NUTT.) NUTT., syn. Brauneria pallida BRITTON, Echinacea angustifolia HOOKER, Echinacea pallida (NUTT.) NUTT. f. albida STEYERM., Rudbeckia pallida NUTT.	Root		Not NFS	×	×		×	

Remarks	Risks	Critical plant substances	Pharmacologically effective dose
	contains coumarin, s. BgVV statement: http://www.bfr.bund.de/cm/ 343/gehalte_an_cumarin_safrol_ methyleugenol_und_estragol_in_ lebensmitteln.pdf	Coumarin	
Positive monograph, indications: paroxysmal and dry cough	Overdose causes inflammations in the mouth and throat, vomiting, ulcers in the digestive tract, bloody diarrhoea (Roth/Daunderer)		**Commission E:** 3 g drug/day, equivalent preparations
Negative monograph: due to risks, toxic II (Wink/Wyk), toxic + (Roth/Daunderer)	Nausea, vomiting, abdominal pain, diarrhoea, cramps, cardiac insufficiency, fainting fits, dyspnoea, respiratory arrest, fatal poisoning (Wink/Wyk, Roth/Daunderer), Commission E: due to the risks (many cases of poisoning, also with fatal outcome), internal use cannot be justified	Acylphloroglucinols, flavaspidic acid	
highly toxic ++ (Roth/Daunderer), very toxic Ib (Wink/Wyk)	Hypersalivation, vomiting, headache, severe, bloody diarrhoea, kidney damage, ingestion of higher doses causes convulsions, death by respiratory arrest (Wink/Wyk)	Cucurbitacins	
Negative monograph: due to risks associated with parenteral use	allergic reactions		**ESCOP:** 3 times 1 ml extract, equivalent preparations
Negative monograph: due to risks associated with parenteral use, **WHO monograph:** Supportive therapy for colds and infections of the respiratory and urinary tract. Beneficial effects in the treatment of these infections are generally thought to be brought about by stimulation of the immune response, **ESCOP monograph:** adjuvant therapy and prophylaxis of recurrent infections of the upper respiratory tract (common cold), **HMPC monograph:** traditional use	allergic reactions		
Negative monograph: due to risks associated with parenteral use	allergic reactions		
Positive monograph, indications: adjuvant treatment of influenza, **WHO monograph:** supportive therapy for colds and infections of the respiratory and urinary tract. Beneficial effects in the treatment of these infections are generally thought to be brought about by stimulation of the immune response, **ESCOP monograph:** adjuvant therapy and prophylaxis of recurrent infections of the upper respiratory tract (common cold), **HMPC monograph:** traditional use	allergic reactions		**Commission E:** tincture equivalent to 900 mg drug/day **ESCOP:** 900 mg drug/day, equivalent preparations **WHO:** 0.5 tsp. (approx. 1 g) in 150 ml water, 3 times a day

Stock plant (Latin)	Plant part	F	NF	MP	trad. MP	List A	List B	List C
Echinacea purpurea (L.) MOENCH, syn. Brauneria purpurea (L.) BRITTON, Echinacea intermedia LINDLEY, Echinacea purpurea (L.) MOENCH f. ligettii STEYERM., Rudbeckia hispida HOFFMGG., Rudbeckia purpurea L.	Herb		Not NFS	×	×		×	
Echinacea purpurea (L.) MOENCH, syn. Brauneria purpurea (L.) BRITTON, Echinacea intermedia LINDLEY, Echinacea purpurea (L.) MOENCH f. ligettii STEYERM., Rudbeckia hispida HOFFMGG., Rudbeckia purpurea L.	Root		Not NFS	×	×		×	
Echinopsis peruviana (BRITTON et ROSE) H.FRIEDRICH et G.D.ROWLEY, syn. Trichocereus peruviana	all plant parts					×		
Elettaria cardamomum (L.) MATON, syn. Alpinia cardamomum ROXB.	Fruit	× S, T		×			×	
Eleutherococcus senticosus (RUPR. et MAXIM.) MAXIM., syn. Acanthopanax senticosus (RUPR. et MAXIM.) HARMS, Hedera senticosa RUPR. et MAXIM.	Root		Not NFS	×	×		×	

Remarks	Risks	Critical plant substances	Pharmacologically effective dose
Positive monograph, indications: adjuvant treatment of recurrent infections of the respiratory tract and the lower urinary tract, **WHO monograph:** supportive therapy for colds and infections of the respiratory and urinary tract, beneficial effects in the treatment of these infections are generally thought to be brought about by stimulation of the immune response, **ESCOP monograph:** adjuvant therapy and prophylaxis of recurrent infections of the upper respiratory tract (common cold) and also of the urogenital tract, **HMPC monograph:** short-term prevention and treatment of coughs and colds	Immediate allergic reactions may occur in individual cases (Commission E monograph)		**Commission E, ESCOP, WHO:** 6–9 ml pressed juice, equivalent preparations **HMPC:** 6–9 ml pressed juice, no administration to children under the age of 12
Negative monograph: due to risks associated Adjuvant therapy and prophylaxis of recurrent infections of the upper respiratory tract (common cold), **HMPC monograph:** traditional use	allergic reactions		**ESCOP:** 3 times 300 mg, equivalent preparations
very toxic Ib (Wink/Wyk), mescaline contained in the drug is a non-marketable narcotic acc. to Annex 1 of the Narcotic Drugs Act (BtMG)	psychotropic, ingestion of higher doses causes gastrointestinal complaints, drop in blood pressure, bradycardia, apnoea, vasodilation, paralysis (Wink/Wyk)	Mescaline	
Positive monograph, indications: digestive problems, **WHO monograph (seeds):** treatment of dyspepsia	Contraindications: in case of gallstones, use only after consultation with a physician (Commission E monograph)		**Commission E, WHO:** 1.5 g drug/day, equivalent preparations
Positive monograph, indications: tonic for strengthening and invigoration in case of fatigue and faintness, decreasing physical performance and powers of concentration as well as in convalescence, **WHO monograph:** prophylactic and restorative tonic for enhancement of mental and physical capacities in cases of weakness, exhaustion and tiredness, and during convalescence, **ESCOP monograph:** decreased mental and physical capacities such as weakness, exhaustion, tiredness and loss of concentration, as well as during convalescence, **HMPC monograph:** traditional use	Contraindications: hypertension (Commission E monograph)		**Commission E, ESCOP, WHO:** 2–3 g drug/day, equivalent preparations

Stock plant (Latin)	Plant part	F	NF	MP	trad. MP	List A	List B	List C
Ephedra spp.	all plant parts			×		×		
Epilobium angustifolium L., syn. Chamaenerion angustifolium SCOP., Epilobium gesneri VILLAIN, Epilobium persicifolium VILL.	Herb	× T						
Epilobium parviflorum SCHRE-BER, syn. Epilobium molle LAM., Epilobium pubescens ROTH, Epilobium rivulare HEGETSCHW., Epilobium villosum CURT.	Herb	× T						
Epimedium grandiflorum MORR., syn. Epimedium macranthum MORR. et DECNE	Herb		×					
Epimedium sagittatum (SIEB. et ZUCC.) MAXIM.	Herb		×					
Equisetum arvense L., syn. Equisetum boreala (L.) BÖRNER, Allosites arvense BRONGN.	Herb	× T		×	×		×	
Erysimum cheiri (L.) CRANTZ, syn. Cheiranthus cheiri L., Cheiranthus corinthius BOISS.	all plant parts					×		
Erythroxylum catuaba A.J. SILVA ex RAYM.-HAMET	Bark		×					
Erythroxylum coca LAM., syn. Erythroxylon coca LAM.	Leaf					×		

Remarks	Risks	Critical plant substances	Pharmacologically effective dose
Positive monograph, indications: respiratory diseases accompanied by mild bronchial spasm in adults and schoolchildren, **WHO monograph:** treatment of nasal congestion due to hay fever, allergic rhinitis, acute coryza, common cold, and sinusitis. The drug is further used as a bronchodilator in the treatment of bronchial asthma (4, 8, 10, 21–23)., toxic II (Wink/Wyk), toxic + (Roth/Daunderer), some of the alkaloids contained in the drug are subject to the provisions of the BtMG (e.g. cathine), Ephedra species and preparations prescription-only, alkaloids contained in the drug (e.g. ephedrine) prescription-only, Ephedra species pharmacy-only. Ephedra is a "scheduled substance" of category 1 acc. to Article 2a of Regulations (EC) No. 273/2004 and No. 111/2005 and is thus subject to the provisions of the Precursors Control Act (GÜG).	sympathomimetic properties similar to amphetamines (Wink/Wyk), insomnia, agitation, irritability, headache, vomiting, micturition difficulties, tachycardia, ingestion of higher doses causes drastic increase in blood pressure, cardiac arrhythmia, development of addiction (Roth/Daunderer), deaths in the USA due to consumption of food supplements containing Ephedra: http://www.bfr.bund.de/de/presseinformation/2002/09/bgvv_und_bfarm_warnen__schwere_gesundheitsschaeden_durch_ephedra_kraut-1007.html	Alkaloids (Ephedra alkaloids, e.g. ephedrine, norpseudoephedrine, cathine)	**Commission E:** 300 mg total alkaloids/day, calculated as ephedrine (special doses for children) **WHO:** 1–6 g drug/day, no administration to children under the age of 6
Positive monograph, indications: posttraumatic and static oedema, irrigation therapy of bacterial or inflammatory diseases of the lower urinary tract and kidney gravel, **WHO monograph:** used internally for kidney and bladder diseases, oedema and as an adjuvant in slimming diets. It is applied as irrigation therapy for infectious and inflammatory diseases of the genitourinary tract, and kidney stones, **HMPC monograph:** traditional use, **standard marketing authorisation: 1239.99.99**	Contraindications: oedemas associated with cardiac or renal insufficiency (Commission E monograph)		**Commission E, WHO:** 6 g drug/day, equivalent preparations
toxic II (Wink/Wyk), toxic + (Roth/Daunderer)	Symptoms of cardiac glycoside poisoning accompanied by pronounced digestive problems (Wink/Wyk)	Cardenolides (e.g. cheirotoxin), glucosinolates	
individually or mixed with other barks designated as catuaba			
very toxic Ib (Wink/Wyk), highly toxic ++ (Roth/Daunderer), psychotropic, subject to the provisions of the Narcotic Drugs Act (BtMG)	psychotropic, strong CNS-stimulant, tachycardia, overdose causes death by respiratory arrest, episodes of schizophrenia, strong psychological addiction (Wink/Wyk)	Alkaloids (e.g. cocaine)	

Stock plant (Latin)	Plant part	F	NF	MP	trad. MP	List A	List B	List C
Eucalyptus globulus LABILL., syn. Eucalyptus cordata MIQ., Eucalyptus diversifolia MIQ., Eucalyptus gigantea DEHNH.	Leaf	× T		×	×		×	
Eucommia ulmoides OLIV.	Bark		Not NFS					×
Euphorbia spp.	all plant parts					×		
Euphrasia officinalis L.	Herb	× T						
Fagopyrum esculentum MOENCH, syn. Fagopyrum cereale (SALISB.) RAFIN.	Fruit	×						
Fagopyrum esculentum MOENCH, syn. Fagopyrum cereale (SALISB.) RAFIN.	Herb	× T		×	×		×	
Feijoa sellowiana BERG, syn. Acca sellowiana (BERG) BURRET, Orthostemon sellowianus BERG	Fruit	×						
Ficus carica L., syn. Caprificus insectifera GASP., Ficus communis LAM., Ficus latifolia SALISB.	Fruit	×						
Filipendula ulmaria (L.) MAXIM, syn. Spiraea ulmaria L., Ulmaria palustris MOENCH, Ulmaria pentapetala GILIB.	Flower, herb	× T		×	×		×	

Remarks	Risks	Critical plant substances	Pharmacologically effective dose
Positive monograph, indications: coughs and colds of the respiratory tract, **standard marketing authorisation: 9299.99.99, WHO monograph (leaves, oil):** traditional use, **ESCOP monograph (oil):** adjuvant treatment of chronic obstructive respiratory complaints including bronchitis and bronchial asthma, symptomatic relief of colds and catarrh of the upper respiratory tract, mildly toxic II to toxic III (Wink/Wyk), slightly toxic + (Roth/Daunderer), **HMPC monograph:** traditional use	Higher doses of the essential oil cause CNS disorders and gastrointestinal complaints (Wink/Wyk), ingestion of higher doses causes drop in blood pressure and bradypnoea, dyspepsia, vomiting, diarrhoea, circulatory distress, cardiac insufficiency, circulatory collapse (Roth/Daunderer), ingestion of Eucalyptus preparations may cause nausea, vomiting and diarrhoea in rare cases, contraindications reported (Commission E monograph)	Eucalyptol (=1,8-cineol)	**ESCOP:** 100–200 mg, 2–5 times a day **WHO:** 4–6 g drug/day, no administration to children
highly toxic ++ (Roth/Daunderer), very toxic Ib (Wink/Wyk)	Irritation of skin and mucous membranes, allergic reactions, gastric pain, nausea and retching, laxative, bloody diarrhoea, arrhythmia, hepatic and renal dysfunction, coma, cocarcinogenic (Wink/Wyk)	Phorbol ester, triterpenes	
Neutral monograph, HMPC assessment report			
peel fruits before consumption	The red hull surrounding the grains contains naphthodianthrones (e.g. fagopyrin). Fagopyrin may induce photodermatitis (fagopyrism) (Ternes)	Naphthodi-anthrones (e.g. fagopyrin)	
mildly toxic III (Wink/Wyk), slightly toxic (+) (Roth/Daunderer), licensed medicinal product	contains naphthodianthrones (e.g. fagopyrin), which may induce photodermatitis. Fagopyrin not detected in aqueous preparations (Frohne), ingestion of high doses causes digestive problems (Wink/Wyk).	Naphthodi-anthrones (e.g. fagopyrin)	
Neutral monograph			
Positive monograph, indications: adjuvant treatment of coughs and colds, **standard marketing authorisation: 1609.99.99, ESCOP monograph:** supportive therapy for the common cold, **HMPC monograph:** traditional use	Contraindications: known hypersensitivity to salicylates (due to the salicylates contained in the drug) (Commission E monograph)		**Commission E:** 2.5–3.5 g flower or 4–5 g herb/day, equivalent preparations **ESCOP:** 2–6 g drug/day

Stock plant (Latin)	Plant part	F	NF	MP	trad. MP	List A	List B	List C
Foeniculum vulgare MILLER ssp. vulgare var. dulce (MILLER) THEULLUNG, syn. Anethum foeniculum dulce SCHKUHR	Fruit	× T, S		×	×		×	
Foeniculum vulgare MILLER ssp. vulgare var. vulgare (MILLER) THELLUNG, syn. Anethum foeniculum vulgare SCHKUHR	Fruit	× T, S		×	×		×	
Fragaria ananassa DUCHESNE ex ROZIER	Leaf	× T						
Fragaria ananassa DUCHESNE ex ROZIER	Fruit	×						
Fragaria vesca L., syn. Fragaria botryformis E. KÖNIG, Fragaria hortensis DUCH.	Leaf	× T						
Fragaria vesca L., syn. Fragaria botryformis E. KÖNIG, Fragaria hortensis DUCH.	Fruit	×						
Fraxinus ornus L., syn. Fraxinus mannifera STEUD., Fraxinus rotundifolia MILL., Ornus europaea PERS.	Resin		Not NFS	×			×	
Fumaria officinalis L., syn. Fumaria media LOIS., Fumaria sturmii OPIZ, Fumaria vulgaris BUB.	Herb		Not NFS	×	×		×	
Galeopsis segetum NECK., syn. Galeopsis ochroleuca LAM.	Herb	× T		×			×	

Remarks	Risks	Critical plant substances	Pharmacologically effective dose
Positive monograph, indications: digestive problems such as mild gastrointestinal cramps, feeling of fullness, bloating, catarrhs of the upper respiratory tract, **standard marketing authorisation:** 5199.99.99, **WHO monograph:** symptomatic treatment of dyspepsia, bloating and flatulence, as an expectorant for mild inflammation of the upper respiratory tract, treatment of pain in scrotal hernia, and dysmenorrhoea, **ESCOP monograph:** dyspeptic complaints such as mild, spasmotic gastrointestinal ailments, bloating and flatulence, catarrh of the upper respiratory tract, included in Annex 1 of Reg. 2001/83 (s. Commission Decision 2008/911/EC), **HMPC monograph:** traditional use	Side effects: allergic reactions of the skin and the respiratory tract observed in individual cases (Commission E monograph)	Trans-anethole, estragole (=methyl chavicol)	**Commission E:** 5–7 g drug/day, equivalent preparations **WHO, ESCOP:** 5–7 g drug/day
Positive monograph, indications: digestive problems such as mild gastrointestinal cramps, feeling of fullness, bloating, catarrhs of the upper respiratory tract, **standard marketing authorisation:** 5199.99.99, **WHO monograph:** symptomatic treatment of dyspepsia, bloating and flatulence, as an expectorant for mild inflammation of the upper respiratory tract, treatment of pain in scrotal hernia, and dysmenorrhoea, **ESCOP monograph:** dyspeptic complaints such as mild, spasmotic gastrointestinal ailments, bloating and flatulence, catarrh of the upper respiratory tract, included in Annex 1 of Reg. 2001/83 (s. Commission Decision 2008/911/EC), **HMPC monograph:** traditional use	Side effects: allergic reactions of skin and respiratory tract observed in individual cases (Commission E monograph)	Trans-anethole, estragole (=methyl chavicol)	**Commission E:** 5–7 g drug/day, equivalent preparations **WHO, ESCOP:** 5–7 g drug/day
Neutral monograph	Hypersensitivity reactions may occur in patients with a predisposition (Schilcher)		
HMPC monograph: in progress			
Positive monograph, indications: constipation, conditions in which easy defaecation with a soft stool is desirable, e.g. anal fissures, haemorrhoids and after rectal-anal surgical procedures	Sensitive patients may experience nausea and bloating, contraindications: intestinal occlusion (Commission E monograph)		**Commission E:** 20–30 g drug/day, equivalent preparations
Positive monograph, indications: cramps in the gall bladder, the biliary tract and the gastrointestinal tract, **standard marketing authorisation:** 1479.99.99, **ESCOP monograph:** digestive complaints (e.g. stomach ache, nausea, vomiting, feeling of fullness, flatulence) due to hepatobiliary disturbance, **HMPC monograph:** traditional use, toxic II (Wink/Wyk), toxic + (Roth/Daunderer)	slightly psychotropic, calmative, narcotic, ingestion of high doses causes burning in the mouth and throat, nausea, vomiting, diarrhoea, drop in blood pressure (Wink/Wyk)		**Commission E:** 6 g drug/day, equivalent preparations **ESCOP:** 4–6 g drug/day, equivalent preparations
Positive monograph, indications: mild catarrhs of the respiratory tract			**Commission E:** 6 g drug, equivalent preparations

Stock plant (Latin)	Plant part	F	NF	MP	trad. MP	List A	List B	List C
Galium odoratum (L.) SCOP., syn. Asperula odorata L., Asperula odora SALISB., Chlorostemma odoratum FOURR.	Herb	× T					×	
Galium verum L., syn. Galium luteum LAM.	Herb	× T						
Garcinia cambogia DESROUSS., syn. Cambogia gutta L.	Fruit	× S						
Gelsemium sempervirens (L.) JAUME ST.-HIL., Syn. Gelsemium nitidum MICHX., Anonymos sempervirens L.	Rhizome			×		×		
Genista tinctoria L., syn. Cystisus tinctorius VIS., Genista depressa BIEB.	Flower					×		
Gentiana lutea L., syn. Asterias lutea BORKH., Swertia lutea VEST.	Root	× T		×	×		×	
Geum urbanum L., syn. Caryophyllata officinalis MOENCH, Caryophyllata urbana SCOP., Caryophyllata vulgaris LAM., Geum caryophyllata GILIB.	Herb		Not NFS					×
Geum urbanum L., syn. Caryophyllata officinalis MOENCH, Caryophyllata urbana SCOP., Caryophyllata vulgaris LAM., Geum caryophyllata GILIB.	Root	× T						

Remarks	Risks	Critical plant substances	Pharmacologically effective dose
Neutral monograph: BfArM graduated plan of 23 May 1997 concerning coumarin-containing phytopharmaceuticals and homeopathic drugs up to potency D2 containing sweet clover (Meliloti herba) or sweet woodruff (Gallii odorati herba) (for internal use), slightly toxic (+) (Roth/Daunderer), maximum amount of coumarin when adding coumarin-containing foods/food ingredients to foods as flavouring specified in Reg. (EC) No. 1334/2008	Consumption of sweet woodruff preparations may cause headache (Roth/Daunderer)	Coumarin	
Negative monograph: extremely toxic Ia (Wink/Wyk), extremely toxic +++ (Roth/Daunderer)	excessive sweating, central paralysis, faintness, convulsions, drop in blood pressure and cardiac activity, visual disturbances, convulsions similar to those induced by strychnine, death by respiratory arrest (Wink/Wyk, Roth/Daunderer)	Indole alkaloids, steroids, iridoid glycosides	
highly toxic ++ (Roth/Daunderer), very toxic Ib (Wink/Wyk)	slightly psychotropic, hallucinogenic, diuretic, uterus contractions, abdominal pain, tachycardia, cardiac arrhythmia, vomiting, diarrhoea, vertigo (Wink/Wyk)	Quinolizidine alkaloids (e.g. cytisine, anagyrine)	
Positive monograph, indications: lack of appetite, digestive problems such as feeling of fullness and bloating, **standard marketing authorisation:** 9199.99.99, **WHO monograph:** treatment of digestive complaints, such as loss of appetite, feeling of distension and flatulence. As an appetite stimulant during convalescence, **ESCOP monograph:** Anorexia e.g after illness, dyspeptic complaints, **HMPC monograph:** traditional use	Headache may occasionally occur in patients with a particular predisposition, contraindications reported (Commission E monograph)		**Commission E:** 2–4 g drug/day **ESCOP:** 0.1–2 g drug/day, special dosage instructions for children **WHO:** 0.1–2 g drug, up to 3 times a day, no administration to children

Stock plant (Latin)	Plant part	F	NF	MP	trad. MP	List A	List B	List C
Ginkgo biloba L., syn. Pterophyllus salisburiensis NELSON, Salisburia adiantifolia SMITH	Leaf	× T		×			×	
Glechoma hederacea L., syn. Calamintha hederacea SCOP.	Leaf	× T						
Gloriosa superba L., syn. Methonica superba LAM. and Gloriosa rothschildiana O'BRIEN. and Gloriosa simplex L., syn. Gloriosa virescens LINDL., Gloriosa plantii LOUD.	all plant parts					×		

Remarks	Risks	Critical plant substances	Pharmacologically effective dose
Negative monograph: due to risks associated with ginkgo leaves as well as various alcoholic extracts due to inadequate evidence of pharmacological effect and risks that cannot be ruled out, **Positive monograph** for standardised acetone/water dry extract, indications: psycho-organic performance disorders, dementia, impaired memory, concentration difficulties, depression, vertigo, tinnitus, headache, **HMPC monograph:** in progress, **WHO monograph:** extracts have been used for symptomatic treatment of mild to moderate cerebrovascular insufficiency (demential syndromes in primary degenerative dementia, vascular dementia, and mixed forms of both) with the following symptoms: memory deficit, disturbance in concentration, depressive emotional condition, dizziness, tinnitus, and headache. Such extracts are also used to improve pain-free walking distance in people with peripheral arterial occlusive disease such as intermittent claudication, Raynaud disease, acrocyanosis, and postphlebitis syndrome, and to treat inner ear disorders such as tinnitus and vertigo of vascular and involutive origine. Extracts and doses other than those described in Dosage forms and Posology are used for similar but milder indications, **ESCOP monograph:** preparations based on standardized extracts: Symptomatic treatment of: mild to moderate demetia symptoms including primary degenerative demtia, vascular dementia and mixed forms; cerebral insufficiency, neurosensory disturbances such as: dizziness/vertigo and tinnitus, enhancement of cognitive performance, symptomatic treatment of peripheral arterial occlusive disease (intermittent claudication), BfArM graduated plan of 27 May 1997 concerning medicinal products containing Ginkgo biloba leaves, **BfR statement:** http://www.bfr.bund.de/cm/343/die_sicherheit_von_ginkgoblaetter_haltigen_tees_kann_wegen_mangelnder_daten_nicht_beurteilt_werden.pdf	Contact allergies may be induced by the ginkgolic acids contained in the drug (Commission E monograph)	Diterpene lactones, ginkgolic acids	**Commission E, ESCOP, WHO:** 120–240 mg extract/day
slightly toxic (+) (Roth/Daunderer)	toxic to animals, cases of poisoning in humans not known (Hager)		
extremely toxic Ia (Wink/Wyk), highly toxic ++ (Roth/Daunderer)	Spindle poison, abortifacient effect, nausea, pain in the throat and oesophagus, cramps, arrhythmia, internal haemorrhage, peripheral neuropathy, abdominal pain, cold extremities, grey skin, tachycardia, respiratory arrest reported (Wink/Wyk, Roth/Daunderer)	Colchicine	

Stock plant (Latin)	Plant part	F	NF	MP	trad. MP	List A	List B	List C
Glycyrrhiza glabra L., syn. Glycyrrhiza echinata LEPECH., Glycyrrhiza glandulifera WALDST. et KIT., Glycyrrhiza hirsuta L., Liquiritia officinalis MOENCH	Root	× T, S		×	×		×	
Gratiola officinalis L.	all plant parts					×		
Grindelia robusta NUTT. and Grindelia squarrosa PURSH (DUN.), syn. Donia squarrosa PURSH, Grindelia squarrosa DUN.	Herb		Not NFS	×	×		×	
Guaiacum officinale L. and Guaiacum sanctum L., syn. Guaiacum verticale ORTEGA	Wood	× F		×			×	
Gymnema silvestre (WILLD.) R. BR., syn. Gymnema parviflorum ROXB., Asclepias geminata ROXB.	Leaf					×		
Gypsophila paniculata L., syn. Arrostia paniculata (L.) RAF., Gypsophila parviflora MOENCH, Saponaria paniculata (L.) NEUMAYER	Root		Not NFS	×			×	
Hamamelis virginiana L., syn. Hamamelis androgyna WALT.	Leaf	× T		×	×		×	

Remarks	Risks	Critical plant substances	Pharmacologically effective dose
Positive monograph, indications: catarrhs of the upper respiratory tract, gastric ulcers, **standard marketing authorisation:** 1309.99.99, **WHO monograph, HMPC monograph:** traditional use, **ESCOP monograph:** adjuvant therapy of gastric and duodenal ulcers and gastritis, coughs and bronchial catarrh, as an expectorant, special precautions as specified in the LMKV [Food Labelling Regulation], as from 13 Dec 2014 as specified in Reg. (EC) No. 1169/2011	Long-term use and higher doses may cause mineral corticoid effects manifesting in sodium and water retention, loss of potassium accompanied by hypertension, oedemas and hypokalaemia, in rare cases also myoglobinurina, contraindications and interactions reported (Commission E monograph)	Glycyrrhizic acid	**Commission E:** liquorice from 5 g/day, Succus Liquiritiae from 0.5 g/day **ESCOP:** 1.5–15 g drug/day **WHO:** 5–15 g drug/day, equivalent preparations
very toxic Ib (Wink/Wyk), highly toxic ++ (Roth/Daunderer)	strong laxative effect, ingestion of high doses may cause digestive problems accompanied by nausea, vomiting, hypersalivation, bloody diarrhoea, nephritis, dyspnoea and cardiac dysfunction, death by respiratory arrest, abortifacient effect (Wink/Wyk)	Cucurbitacins	
Positive monograph, indications: catarrhs of the upper respiratory tract, **ESCOP monograph:** productive cough, catarrhs of the upper respiratory tract, **HMPC monograph:** traditional use	Side effects: irritation of gastric mucosa observed in rare cases (Commission E monograph)		**Commission E, ESCOP:** 4–6 g drug/day, equivalent preparations
Positive monograph, indications: adjuvant treatment of rheumatic conditions, mildly toxic II to toxic III (Wink/Wyk), toxic + (Roth/Daunderer)	Ingestion of higher doses causes gastroenteritis accompanied by nausea and diarrhoea, cardiac palpitation, vertigo (Wink/Wyk)		**Commission E:** 4.5 g drug, equivalent preparations
toxic II (Wink/Wyk), toxic + (Roth/Daunderer)	Lack of appetite, faintness, fever, diarrhoea, death by apnoea, irritation of cardiovascular system, polyuria, irritation of uterus, hypoglycaemia (Roth/Daunderer) (Wink/Wyk)	Gymnemic acid	
Positive monograph, indications: catarrhs of the upper respiratory tract	Irritation of gastric mucosa observed in rare cases (Commission E monograph)		**Commission E:** 30–150 mg drug/day, equivalent preparations
Positive monograph, indications: superficial skin lesions, local inflammations of skin and mucosa, haemorrhoids, varicose veins, **standard marketing authorisation:** 9699.99.99 (witch hazel leaves), 9799.99.99 (witch hazel bark), **WHO monograph:** external use only, **ESCOP monograph:** symptomatic treatment of complaints related to varicose veins such as painful and heavy legs and of haemorrhoids, **HMPC monograph:** traditional use		Tanning agents (tannins)	**Commission E:** 0.1–1 g drug, several times a day, equivalent preparations **ESCOP:** 2–3 g drug 2–3 times a day, equivalent preparations **WHO:** 0.1–1 g drug/day, equivalent preparations

Stock plant (Latin)	Plant part	F	NF	MP	trad. MP	List A	List B	List C
Hamamelis virginiana L., syn. Hamamelis androgyna WALT.	Bark		Not NFS	×	×		×	
Harpagophytum procumbens (BURCH.) DC., syn. Harpagophytum burcherllii DECNE	Root	× T		×	×		×	
Harungana madagascariensis LAM. ex POIR., syn. Haronga madagascariensis (LAM. ex POIR.) CHOISY	Leaf, bark		Not NFS	×			×	

Remarks	Risks	Critical plant substances	Pharmacologically effective dose
Positive monograph, indications: superficial skin lesions, local inflammations of skin and mucosa, haemorrhoids, varicose veins, **standard marketing authorisation:** 9699.99.99 (witch hazel leaves), 9799.99.99 (witch hazel bark), **WHO monograph:** external use only, **ESCOP monograph:** inflammation of mucous membranes of the oral cavity, short-term symptomatic treatment of diarrhoea, **HMPC monograph:** traditional use		Tanning agents (tannins)	**Commission E:** 0.1–1 g drug, several times a day, equivalent preparations **ESCOP:** 2–3 g drug 2–3 times a day, equivalent preparations **WHO:** 0.1–1 g drug/day, equivalent preparations
Positive monograph, indications: lack of appetite, digestive problems, adjuvant treatment of degenerative diseases of the locomotor system, **WHO monograph:** treatment of pain associated with rheumatic conditions, treatment of loss of appetite and dyspeptic complaints, supportive treatment of degenerative rheumatism, painful arthrosis and tendonitis, **ESCOP monograph:** symptomatic treatment of painful osteoarthritis, relief of low back pain, loss of appetite and dyspepsia, **HMPC monograph:** traditional use	Contraindications: gastric and duodenal ulcers, in case of gallstones, use only after consultation with a physician (Commission E monograph)		**Commission E:** 1.5 g/day, preparations of equivalent bitter value (lack of appetite), otherwise 4.5 g drug/day, equivalent preparations, standardised extracts equivalent to 30–100 mg harpagoside **ESCOP:** 2–5 g drug/day, equivalent preparations **WHO:** 1–3 g drug/day, equivalent preparations
Positive monograph, indications: digestive problems, mild exocrine pancreatic insufficiency	Photosensitisation especially in light-skinned patients due to the hypericin and pseudohypericin contained in the drug, contraindications reported (Schilcher)	Naphthodianthrone derivatives (leaves: hypericin, pseudohypericin), anthraquinones (bark: madagascin)	**Commission E:** 25–50 mg drug/day, equivalent preparations

Stock plant (Latin)	Plant part	F	NF	MP	trad. MP	List A	List B	List C
Hedera helix L., syn. Hedera caucasigena POJARK, Hedera chrysocarpa WALSH	Leaf			×	×	×		
Helianthus annuus L., syn. Helianthus cultus VENTSL., Helianthus erythrocarpus BARTL., Helianthus indicus L., Helianthus pumilus PERS.	Petal	× T						
Helianthus annuus L., syn. Helianthus cultus VENTSL., Helianthus erythrocarpus BARTL., Helianthus indicus L., Helianthus pumilus PERS.	Seed	×						
Helianthus tuberosus L., syn. Helianthus mollissimus E.E. WATSON	Stem tuber	×						
Helichrysum arenarium (L.) MOENCH, syn. Gnaphalium arenarium L.	Flower	× T		×			×	

Remarks	Risks	Critical plant substances	Pharmacologically effective dose
Positive monograph, indications: catarrhs of the upper respiratory tract, symptomatic treatment of chronic bronchitis, **ESCOP monograph:** coughs, particularly when associated with hypersecretion of viscous mucus, as adjuvant treatment of inflammatory bronchial diseases, toxic II (Wink/Wyk), toxic + (Roth/Daunderer), toxic + (Roth/Daunderer), **HMPC monograph:** expectorant for productive cough	Local irritations of mucous membranes, digestive tract irritation, dizziness, tachycardia, headache, nausea, vomiting, skin rash, CNS disorders, hyperthermia and cardiac palpitation observed in individual cases, especially in higher doses, overdose may cause death by respiratory arrest (Wink/Wyk, Roth/Daunderer)	Saponins	**Commission E:** average daily dose: 0.3 g drug, equivalent preparations **ESCOP:** ethanolic preparations: 250–420 mg drug/day, deviating doses for children, ethanol-free preparations: 300–945 mg drug/day, deviating doses for children **HMPC: adults and adolescents:** a) single dose: 15–65 mg, 1–3 times a day up to a daily dose of 45–105 mg (note: max. daily dose for ethanolic ready-to-use products: 67 mg equivalent to 420 mg drug) b) single dose: 14–18 mg, 3 times a day c) single dose: 33 mg, 2 times a day d) single dose: 100 mg, 3 times a day **children aged between 6 and 12:** a) single dose: 11–33 mg, 2–3 times a day up to a max. daily dose of 33–70 mg (note: max. daily dose for ethanolic ready-to-use products: 34 mg equivalent to 210 mg drug) b) single dose: 9–18 mg, 2–3 times a day up to a max. daily dose of 15–40 mg c) single dose 25 mg, 2 times a day d) single dose: 75 mg 3 times a day **children aged between 2 and 5:** a) single dose: 8–18 mg 2–3 times a day up to a max. daily dose of 24–36 mg (note: max. daily dose for ethanolic ready-to-use products: 24 mg equivalent to 150 mg drug) **No administration to children under the age of 2**
Positive monograph, indications: digestive problems, **WHO monograph:** treatment of dyspeptic disorders, **standard marketing authorisation:** 1649.99.99	Contraindications: occlusion of biliary tract, in case of gallstones, use only after consultation with a physician (Commission E monograph)		**Commission E:** 3 g drug/day, equivalent preparations

Stock plant (Latin)	Plant part	F	NF	MP	trad. MP	List A	List B	List C
Heliotropium arborescens L., syn. Heliotropium corymbosum RUIZ et PAV, Heliotropium odoratum MOENCH, Heliotropium peruvianum L.	all plant parts					×		
Helleborus spp.	all plant parts			×		×		
Herniaria glabra L.	Herb	× T			×			
Herniaria hirsuta L.	Herb	× T			×			
Hibiscus sabdariffa L., syn. Sabdariffa rubra KOSTEL., Sida sabdariffa L.	Flower	× T						
Hibiscus sabdariffa L., syn. Sabdariffa rubra KOSTEL., Sida sabdariffa L.	Seed		Not NFS					×
Hippophae rhamnoides L. ssp. rhamnoides, Hippophae rhamnoides L. ssp. fluviatilis, syn. Hippophae angustifolia "LODD." ex DIPPEL, Hippophae fasciculata WALL ex STEUDEL, Hippophae littoralis SALISB.	Fruit	×		×			×	
Hoodia gordonii (MASS.) SWEET	all plant parts		×					
Hordeum vulgare L., syn. Hordeum sativum JESS.	Fruit	×						
Humulus lupulus L., syn. Humulus cordifolius MIQ.	Flower	×		×	×		×	
Hydrastis canadensis L., syn. Warneria canadensis LILL., Warneria diphylla RAF.	Rhizome			×		×		
Hyoscyamus spp.	all plant parts			×		×		

Remarks	Risks	Critical plant substances	Pharmacologically effective dose
toxic II (Wink/Wyk), toxic + (Roth/Daunderer), substance of concern (AMK), BfArM graduated plan of 5 June 1992 concerning medicinal products containing pyrrolizidine alkaloids	Pyrrolizidine alkaloids contained in the drug may cause liver damage (Wink/Wyk, Frohne), tachycardia, mydriasis, central apnoea and cardiac arrest (Wink/Wyk).	Pyrrolizidine alkaloids	
Helleborus viridis, Helleborus niger and various species pharmacy-only, extremely toxic +++ (Roth/Daunderer), extremely toxic Ia (Wink/Wyk)	tickle in mouth and throat, nausea, hypersalivation, mydriasis, raging thirst, abdominal pain and colic, gastrointestinal complaints, diarrhoea, exhaustion, symptoms of cardiac glycoside poisoning, coma, death by cardiac or respiratory arrest (Wink/Wyk)	Alkaloids (e.g. celliamine, sprintillamine), cardioactive glycosides (e.g. bufa-dienolides), steroidal saponins (e.g. helleborine)	
Neutral monograph			
Neutral monograph			
Neutral monograph			
WHO monograph: treatment of cirrhosis of the liver, used to relieve cough with profuse expectoration, to promote digestion in people with prolonged gastrointestinal transit with abdominal pain, and for treatment of amenorrhoea			WHO monograph: 40–50 g fresh fruit in 1 litre water, 2–3 cups a day
Positive monograph, indications: emotional symptoms such as agitation and states of anxiety, sleep disturbances, standard marketing authorisation: 1029.99.99, WHO monograph: as a sedative for the treatment of nervous tension and insomnia. Treatment of dyspepsia and lack of appetite, ESCOP monograph: tenseness, restlessness and sleep disorders, HMPC monograph: traditional use	fresh hop cones: dermatitis on skin contact, conjunctivitis, in severe cases vomiting, fever, dyspnoea, bradycardia and sweats (Roth/Daunderer)		Commission E, ESCOP, WHO: single dose: 0.5 g drug, equivalent preparations, no administration to children under the age of 12
WHO monograph: treatment of digestive complaints, such as dyspepsia, gastritis, feeling of distension and flatulence, toxic Ib to very toxic II (Wink/Wyk), highly toxic ++ (Roth/Daunderer)	Ingestion of higher doses causes vomiting, gastrointestinal complaints, uterus contractions, vasoconstrictive, CNS-depressor, spasmogenic, may cause myoplegia, hallucinations, delirium and cyanosis (Wink/Wyk)	Isoquino-line alkaloids (e.g. hydrastine, berberine)	WHO: 0.5–1 g drug, 3 times a day
Positive monograph for Hyoscyamus niger, indications: gastrointestinal cramps, extremely toxic Ia (Wink/Wyk), extremely toxic +++ (Roth/Daunderer)	psychotropic, calmative and sedative in low doses, ingestion of higher doses causes hallucinations, euphoria, disorientation, insomnia, fits of rage, loss of consciousness and death by respiratory arrest, facial blush, dry mouth, mydriasis, accelerated pulse (Wink/Wyk)	Tropane alkaloids (e.g. L-hyoscyamine, L-scopolamine, atropine)	Commission E: 0.5 g standardised powder (single dose), 3.0 g standardised powder (maximum daily dose)

Stock plant (Latin)	Plant part	F	NF	MP	trad. MP	List A	List B	List C
Hypericum perforatum L., syn. Hypericum officinarum CRANTZ, Hypericum officinale GATER ex STEUD., Hypericum vulgare LAM.	Herb, flower	× F		×	×		×	
Hyssopus officinalis L., syn. Hyssopus arisatus GODR., Hyssopus canascens (DC.) NYM.	all plant parts	× S, T						
Iberis amara L., syn. Biauricula amara BUBANI, Crucifera iberis KRAUSE, Thlaspi amarum CRANTZ	all plant parts			×		×		
Ilex aquifolium L.	Fruit, leaf					×		
Ilex paraguariensis ST. HIL., syn. Ilex bonplandiana MÜNTER, Ilex domestica REISS.	Leaf	× T		×	×		×	
Illicium verum HOOK. FIL., syn. Illicium anisatum GAERTN. NON. L., Illicium stellatum L.	Fruit	× T, S		×			×	
Inula helenium L., syn. Aster helenium (L.) SCOP., Corvisartia helenium (L.) MÉRAT, Helenium grandiflorum GILIB.	Root	× S, T			×			
Ipomoea purga (WENDER.) HAYNE, syn. Exogonium purga (WENDER.) BENTH.	all plant parts			×		×		

Remarks	Risks	Critical plant substances	Pharmacologically effective dose
Positive monograph, indications: psychovegetative disorders, depression, anxiety and/or nervous agitation. Oily Hypericum preparations for digestive problems, **standard marketing authorisation:** 1059.99.99, **WHO monograph:** symptomatic treatment of mild and moderate depressive episodes (classified as F32.0 and F32.1, respectively, in the International statistical classification of diseases and related health problems, Tenth revision (ICD-10), **ESCOP monograph:** episodes of mild depressive disorders or mild to moderate depressive episodes in accordance with ICD-10 categories F32.0, F32.1, F33.0 and F33.1, **HMPC monograph:** herbal medicinal product for the treatment of mild to moderate depressive episodes (according to ICD-10). short term treatment of symptoms in mild depressive disorders, BfArM graduated plan of 10 Oct 2005 concerning medicinal products for internal use containing St. John's wort (Hypericum): http://www.bfarm.de/SharedDocs/1_Downloads/DE/Pharmakovigilanz/stufenplverf/Johanniskraut-Bescheid-051010.pdf?_	Photosensitiser, enzyme inductor (suppression of effect of concomitant medication), toxic increase of serotonin concentration in CNS (Wink/Wyk)	Naphthodianthrones (e.g. hypericin, pseudohypericin), xanthone derivatives	**Commission E:** 2–4 g drug/day **ESCOP, WHO:** 2–4 g drug/day, administration to children only under medical supervision
Neutral monograph: use of hyssop as taste improver in tea blends in amounts of up to 5 % is considered to be safe (Commission E monograph)	Consumption of oil reported to cause clonic and/or tonic-clonic cramps (Commission E monograph)	Pinocamphone, esopinocamphone	
Ingredient of licensed medicinal products, toxic + (Roth/Daunderer), mildly toxic II to toxic III (Wink/Wyk)	Ingestion of higher doses causes vomiting, gastroenteritis, bloody diarrhoea, abdominal pain, respiratory depression, arrhythmia (Wink/Wyk)	Cucurbitacins	
highly toxic ++ (Roth/Daunderer), toxic Ib to very toxic II (Wink/Wyk)	gastrointestinal complaints, gastritis, vomiting, diarrhoea, cardiac arrhythmia, paralysis, kidney damage, drowsiness, possibly death (Wink/Wyk)	Triterpenes, nitrile (menesdaurin)	
Positive monograph, indications: mental and physical fatigue, **HMPC monograph:** traditional use			**Commission E:** 3 g drug/day, equivalent preparations
Positive monograph, indications: catarrhs of the respiratory tract, digestive problems, **standard marketing authorisation:** 2419.99.99			**Commission E:** 3 g drug or 0.3 g essential oil/day, equivalent preparations
Negative monograph: due to risks, practically non-toxic but allergenic (Roth/Daunderer)	Sesquiterpene lactones contained in elecampane are irritating to mucous membranes, sensitising and induce allergic contact dermatitis (Commission E monograph), ingestion of higher doses causes vomiting, diarrhoea, cramps, symptoms of paralysis (Roth/Daunderer)	Sesquiterpene lactones (alantolactones, isoalantolactones)	
toxic II (Wink/Wyk), toxic (Roth/Daunderer), resin and root tuber as MP prescription-only, whole plant pharmacy-only	drastic laxative, gastric pain, vomiting, diarrhoea, tenesmus, nausea (Wink/Wyk, Roth/Daunderer)	Convolvulin, jalapin, glycoretin	

Stock plant (Latin)	Plant part	F	NF	MP	trad. MP	List A	List B	List C
Ipomoea violacea L., syn. Ipomoea tricolor CAV.	Seed					×		
Iris germanica L. and **Iris pallida LAM.** and **Iris florentina L.**	Rhizome	× F						
Iris versicolor L., syn. Iris virginica JACQ., PURSH (non L.)	Root		×					
Jasminum grandiflorum L.	Flower	× T						
Jasminum officinale L.,	Flower	× T						
Juglans regia L., syn. Nux juglans DUHAM.	Flower	× C					×	
Juglans regia L., syn. Nux juglans DUHAM.	Seed	×						
Juglans regia L., syn. Nux juglans DUHAM.	Peel	× F						
Juniperus communis L.	Fruit	× S, T		×	×		×	
Juniperus communis L.	Wood	× T						
Juniperus sabina L., syn. Sabina officinalis GARCKE	all plant parts			×		×		
Krameria triandra RUIZ et PAVON, syn. Krameria lappacea (DOMBEY) BURDET et SIMPSON, Krameria canascens WILLD. ex SCHULTES	Root	× T						
Laburnum anagyroides MEDICUS, syn. Cytisus laburnum L., Genista laburnum KRAUSE	all plant parts			×		×		

Remarks	Risks	Critical plant substances	Pharmacologically effective dose
highly toxic ++ (Roth/Daunderer), very toxic Ib (Wink/Wyk), narcotic	psychotropic, hallucinogenic effect, nausea, vomiting and faintness, ingestion of high doses causes death by apnoea (Roth/Daunderer, Wink/Wyk)	Indole alkaloids (e.g. ergometrine, ergine, chanoclavin, lysergic acid and lysergic acid derivatives)	
Neutral monograph: used to produce spirits			
Positive monograph: for external use, **standard marketing authorisation:** 2429.99.99, **HMPC monograph:** traditional use (external use)	The high tannin content may cause irritation of gastric mucosa. Since a mutagenic effect is demonstrated for juglone, internal use of the drug is contraindicated in pregnancy (Hager)	Juglone	
Negative monograph: due to risks, as flavouring in spirits production	Local daily use of preparations containing juglone is associated with a high incidence of tongue cancer and leucoplakia (Commission E monograph)	Juglone	
Positive monograph, indications: digestive problems, **standard marketing authorisation:** 1369.99.99, **ESCOP monograph:** traditional use, non-toxic to slightly toxic (+) (Roth/Daunderer), mildly toxic II to toxic III (Wink/Wyk), **HMPC monograph:** traditional use	Long-term use or overdose (more than 150 mg of essential juniper oil per day, note by Schilcher) may cause kidney damage (Schilcher), contraindications reported (Commission E monograph)	Mono- and sesquiterpenes (e.g. alpha-pinene)	**Commission E:** 2–10 g/day equivalent to 20–100 mg essential oil **ESCOP:** 2–3 g drug in 150 ml water, 3–4 times a day
		Terpene, tanning agents, lignans	
extremely toxic +++ (Roth/Daunderer), extremely toxic Ia (Wink/Wyk), substance of concern (AMK)	Ingestion causes nausea, vomiting, abdominal pain, excitement, arrhythmia, tachycardia, convulsions, respiratory arrest, serious kidney and liver damage, profuse internal haemorrhage, haematuria, uterus contractions, abortion, death as a result of central apnoea in coma (Wink/Wyk). Essential oil has a strong irritating effect on the skin, which may even cause necrosis and damage to deeper skin layers.	Monoterpenes (e.g. thujone, sabinene)	
Positive monograph: for external use, **standard marketing authorisation:** 1179.99.99 (root), 7199.99.99 for external or local use only, **ESCOP monograph:** for external or local use	allergic reactions of mucous membranes observed in very rare cases (Commission E monograph)		**ESCOP, Commission E:** external use only
very toxic Ib (Wink/Wyk), extremely toxic +++ (Roth/Daunderer)	slightly psychotropic and hallucinogenic, diuretic, uterus contractions, abdominal pain, tachycardia, drop in blood pressure, cardiac arrhythmia, vomiting, diarrhoea, vertigo, cramps, headache, delirium, nicotinergic effect (Wink/Wyk)	Quinolizidine alkaloids (e.g. cytisine)	

Stock plant (Latin)	Plant part	F	NF	MP	trad. MP	List A	List B	List C
Lactuca virosa L., syn. Lactuca altissima M.B., Lactuca lactucarii LAMOTTE, Lactuca sylvestris LAM.	all plant parts			×		×		
Lamium album L., syn. Lamium capitatum SM., Lamium vulgatum var. album BENTH	Flower	× T		×			×	
Lamium album L., syn. Lamium capitatum SM., Lamium vulgatum var. album BENTH	Herb	× T						
Laurus nobilis L.	Leaf	× S						
Lavandula angustifolia MILL., syn. Lavandula officinalis CHAIX ex VILL., Lavandula angustifolia MUCH, Lavandula spica L.	Leaf	× T		×	×		×	
Ledum palustre L., syn. Rhododen-dron tomentosum HARMAJA	Herb					×		
Leonurus cardiaca L., syn. Cardiaca trilobata LAM., Cardiaca vulgaris MOENCH	Herb	× T		×	×		×	
Lepidium peruvianum CHACON, syn. Lepidium meyenii WALP.	Root	×						
Lespedeza capitata MICHX.	all plant parts		Not NFS					×
Levisticum officinale KOCH, syn. Angelica levisticum ALL., Hip-poselinum levisticum BRITTON et ROSE, Levisticum levisticum KARSTEN	Fruit	× S						

Remarks	Risks	Critical plant substances	Pharmacologically effective dose
very toxic Ib (Wink/Wyk), highly toxic ++ (Roth/Daunderer)	persistent irritation of skin and mucous membranes, calmative properties similar to opium, visual disturbances, headache, sweats, vertigo, drowsiness, tachypnoea and tachycardia (Wink/Wyk, Roth/Daunderer)	Sesquiterpene lactones	
Positive monograph, indications: catarrhs of the upper respiratory tract, mild infections of the oral and pharyngeal mucosa			**Commission E:** 3 g drug/day, equivalent preparations
Neutral monograph			
	irritating to skin, allergenic (Roth/Daunderer)	Eucalyptol (=1,8-cineol), sesquiterpene lactones	
Positive monograph, indications: emotional symptoms such as agitation, sleep disturbances, functional epigastric disorders (nervous irritable stomach, ROEMHELD syndrome, meteorism, nervous intestinal disorders), **standard marketing authorisation:** 1119.99.99, **WHO monograph:** symptomatic treatment of restlessness, insomnia, and as a carminative and antispasmodic for gastrointestinal disorders of nervous origin, **ESCOP monograph:** symptoms of mood disturbance such as restlessness, agitation of insomnia, functional abdominal complaints, mildly toxic III (Wink/Wyk), **HMPC monograph:** traditional use	Overdose may induce hepatic and renal dysfunction (Wink/Wyk)	Monoterpenes, coumarins	**Commission E, ESCOP, WHO:** 1–2 tsp. Drug, 0.8–1.6 g/cup (150 ml), 3 times a day
Negative monograph: due to risks, toxic (Roth/Daunderer), toxic II (Wink/Wyk)	gastrointestinal irritation accompanied by vomiting and diarrhoea, irritation and damage of kidneys and lower urinary tract, sweats, pain in muscles and joints, CNS excitement accompanied by states of intoxication, cramps and subsequent paralysis (Commission E monograph)	Sesquiterpenes (e.g. ledol, palustrol), arbutin	
Positive monograph, indications: nervous cardiac disorders, adjuvant treatment of hyperthyroidism, **WHO monograph:** used against cerebral ischaemia, it is also used for treatment of heart palpitations occurring with anxiety attacks or other nervous disorders, **HMPC monograph:** traditional use, mildly toxic III (Wink/Wyk)	slightly anaesthetic, calmative (Wink/Wyk)	Iridoids, diterpenes, triterpenes	**Commission E:** 4.5 g drug, equivalent preparations **WHO:** 2–4.5 g drug, equivalent preparations
Statement of the Federal Institute for Risk Assessment (BfR): http://www.bfr.bund.de/cm/343/risikobewertung_macahaltiger_nahrungsergaenzungsmittel.pdf		Glucosinolates, imidazole alkaloids	
	phototoxic	Furocoumarins	

Stock plant (Latin)	Plant part	F	NF	MP	trad. MP	List A	List B	List C
Levisticum officinale KOCH, syn. Angelica levisticum ALL., Hipposelinum levisticum BRITTON et ROSE, Levisticum levisticum KARSTEN	Herb	× S						
Levisticum officinale KOCH, syn. Angelica levisticum ALL., Hipposelinum levisticum BRITTON et ROSE, Levisticum levisticum KARSTEN	Root	× S, T		×	×		×	
Linaria vulgaris MILL., syn. Antirrhinum linaria L.	Herb		Not NFS					×
Linum usitatissimum L.	Seed	×		×	×		×	
Lippia triphylla (L'HÉR.) KUNTZE, syn. Aloysia citrodora PALAU, Aloysia citriodora ORTEGA ex PERS., Aloysia sleumeri MOLDENKE, Aloysia triphylla (L'HÉR.) KUNTZE, Verbena citriodora CAV., Verbena triphylla L'HÉR., Zapania citrodora LAM.	Herb	× S, T						
Lobelia spp.	all plant parts			×		×		
Lophophora williamsii (LEM. ex SALM-DYCK) COULT., syn. Anhalonium williamsii (LEM. ex SALM-DYCK) LEM., Anhalonium lewinii HENN., Ariocarpus williamsii VOSS., Echinocactus lewinii (HENN.) SCHUM.	all plant parts					×		
Lupinus spp.	Seed	×						

Remarks	Risks	Critical plant substances	Pharmacologically effective dose
	phototoxic	Furocoumarins	
Positive monograph, indications: irrigation therapy of inflammatory diseases of the lower urinary tract, irrigation therapy to prevent of kidney gravel, **HMPC monograph:** traditional use, **standard marketing authorisation:** 1569.99.99	phototoxic	Furocoumarins	**Commission E:** 4–8 g drug/day, equivalent preparations
Positive monograph, indications: habitual constipation, colon affected by laxatives, irritable colon, diverticulitis, as gruel preparation for gastritis and enteritis, **standard marketing authorisation:** 1099.99.99, **ESCOP monograph:** constipation, irritable bowel syndrome, diverticular disease, symptomatic short-term treatment of gastritis and enteritis, **HMPC monograph:** treatment of habitual constipation or in conditions in which easy defaecation with soft stool is desirable, traditional use, toxic II (Wink/Wyk)	Cases of poisoning in humans caused by cyanogenic glycosides in flax not known to date (Hager, Frohne/ Pfänder)	cyanogenic glycosides, lignans	**Commission E:** 1 tbsp./150 ml liquid, 2–3 times a day **ESCOP:** 5–10 g drug/day **HMPC:** 10–15 g, 2–3 times a day, no administration to children under the age of 12
Lobelia inflata prescription-only, Lobelia species in general pharmacy-only, Lobelia inflata: highly toxic ++ (Roth/Daunderer), very toxic Ib (Wink/Wyk)	nicotinergic effect, vomiting, diarrhoea accompanied by colic and cramps, stimulates respiration in low doses, causes apnoea in higher doses, bradycardia, drop in blood pressure, convulsions, death by respiratory arrest (Wink/Wyk, Roth/Daunderer)	Piperidine alkaloids (e.g. lobeline)	
narcotic, highly toxic ++ (Roth/Daunderer), very toxic Ib (Wink/Wyk), mescaline contained in the drug is a non-marketable narcotic acc. to Annex 1 of the Narcotic Drugs Act (BtMG)	psychotropic, hallucinogenic, visual disturbances and distorted thinking patterns, intense, even schizophrenic symptoms, loss of perception, insomnia, drop in blood pressure, mydriasis, pain, respiratory depression, vasodilation, faintness, liver damage and vomiting, ascending paralysis in high doses (Wink/Wyk, Roth/Daunderer)	Alkaloids (e.g. mescaline)	
low-alkaloid/alkaloid-free types are used in foods (sweet lupine or leached, ripe seeds of bitter lupines) (van Wyk), other types moderately to highly toxic (Roth/Daunderer), toxic (Wink/Wyk), subject to labelling requirement acc. to Food Labelling Regulation, statement of the Federal Institute for Risk Assessment (BfR) on the allergenicity of lupine proteins: http://www.bfr.bund.de/cm/343/allergie-durch-lupineneiweiss-in-lebensmitteln.pdf	Lupine poisoning (caused by alkaloids) may cause hypersalivation, vomiting, swallowing difficulties, arrhythmia, convulsions and death by respiratory arrest (Wink/Wyk)	Quinolizidine alkaloids	

Stock plant (Latin)	Plant part	F	NF	MP	trad. MP	List A	List B	List C
Lycium barbarum L., syn. Lycium halimifolium MILLER, Lycium flaccidum (VEILLARD) K. KOCH	Fruit	×						
Lycopodium clavatum L., syn. Lepidotis clavata (L.) PAL. BEAUV.	Herb					×		
Lycopus europaeus L., syn. Lycopus aquaticus MOENCH, Lycopus palustris LAM., Lycopus vulgaris PERS. and Lycopus virginicus L., syn. Lycopus macrophyllus BENTH., Lycopus uniflorus MICHX., Lycopus virginicus MICHX.	Herb			×		×		
Macadamia ternifolia F. v. MUELL.	Seed	×						
Malpighia punicifolia L., syn. Malpighia glabra L.	Fruit	×						
Malus domestica BORKH.	Fruit	×						
Malus sylvestris (L) MILL.	Fruit	×						
Malva sylvestris L., syn. Malva ambigua GUSS., Malva erecta C. PRESL., Malva mauritiana L.	Leaf, flower	× T		×			×	
Mandragora officinarum L.	Root			×		×		
Mangifera indica L.	Fruit	×						
Marrubium vulgare L., Marrubium album GILIB.	Herb	×		×	×		×	
Marsdenia condurango REICHB. F., syn. Gonolobus condurango (REICHB. F.) TRIANA	Bark	× T		×	×		×	

Remarks	Risks	Critical plant substances	Pharmacologically effective dose
mildly toxic III (Wink/Wyk), highly toxic ++ (Roth/Daunderer), the presence of toxic amounts of tropane alkaloids, which is the basis of the above assessment, is not confirmed in a recent BfR statement on L. barbarum: http://www.bfr.bund.de/cm/350/risikobewertung-von-pflanzen-und-pflanzlichen-zubereitungen.pdf			
toxic II (Wink/Wyk), toxic + (Roth/Daunderer)	neurotoxic, symptoms of poisoning similar to those induced by curare, excessive sweating, nausea, vertigo, psychotropic, causes impaired sensory perception, impaired speech, cramps, vomiting, diarrhoea (Wink/Wyk)	Quinolizidine alkaloids (e.g. lycopodine, clavatin)	
Positive monograph, indications: mild forms of hyperthyroidism accompanied by vegetative-nervous disorders, mastalgia	Long-term use and high doses may cause enlargement of the thyroid gland, sudden discontinuation of treatment intensifies the symptoms, interactions and contraindications reported (Commission E monograph)	Coumarins	**Commission E:** 1–2 g drug/day (tea), water-ethanol extract equivalent to 20 mg drug
Positive monograph, indications: irritation of oral and pharyngeal mucosa accompanied by dry cough, **standard marketing authorisation:** 1579.99.99 (mallow leaves), **ESCOP monograph for flowers:** dry cough, irritation of the oral, pharyngeal or gastric mucosa			**Commission E:** 5 g drug/day, equivalent preparations **ESCOP:** 1.5–2 g drug/day
psychotropic, root and preparations thereof prescription-only, mandrake pharmacy-only, very toxic Ib (Wink/Wyk), highly toxic ++	calmative and sedative in low doses, ingestion of higher doses causes hallucinations, euphoria, disorientation, insomnia, vertigo, tremor, impaired speech, excitement, nausea, vomiting, mydriasis and death by central paralysis (Wink/Wyk)	Tropane alkaloids (e.g. L-hyoscyamine, D,L-scopolamine, atropine)	
Positive monograph, indications: lack of appetite, digestive problems such as feeling of fullness and bloating, catarrhs of the respiratory tract, slightly toxic (+) (Roth/Daunderer), **HMPC monograph:** traditional use	Ingestion of high doses causes cardiac arrhythmia (Roth/Daunderer)	Diterpenes (e.g. marrubiin)	**Commission E:** 4.5 g drug/day, 2–6 tbsp. pressed juice, equivalent preparations
Positive monograph, indications: lack of appetite, mildly toxic II to toxic III (Wink/Wyk), slightly toxic (+) (Roth/Daunderer)	Ingestion of high doses may cause digestive problems and CNS disorders (Wink/Wyk), hypersalivation, vomiting, stiffness of extremities, convulsive fits accompanied by twitching of facial muscles, central apnoea (Roth/Daunderer)	steroidal glycosides (e.g. condurangins)	**Commission E:** 2–4 g drug/day, equivalent preparations

Stock plant (Latin)	Plant part	F	NF	MP	trad. MP	List A	List B	List C
Matricaria recutita L., syn. Chamomilla recutita (L.) RAUSCHERT, Matricaria chamomilla L. PROPARTE	Flower	× T		×			×	
Matricaria recutita L., syn. Chamomilla recutita (L.) RAUSCHERT, Matricaria chamomilla L. PROPARTE	Seed		Not NFS					×
Medicago sativa L.	Sprout	×						
Melaleuca alternifolia (MAIDEN and BETCH) CHEEL., syn. Melaleuca linariifolia var. alternifolia MAIDEN et BETCHE.	Leaf		×					
Melilotus officinalis (L.) PALLAS, syn. Melilotus arvensis WALLR., Melilotus melilotus-officinalis AS-CHERS. et GRAEBN., Melilotus officinalis DESR. and Melilotus altissima THUILL., syn. Melilotus gigantea WENDER., Melilotus macrorrhiza PERS.	Herb	× T		×	×		×	
Melissa officinalis L., Melissa altissima SM. INSIBTH. et SM., Thymus melissa E.H.L. KRAUSE	Leaf	× S, T		×	×		×	
Mentha aquatica L., syn. Mentha hirsuta HUDS.	Herb	× S, T						

Remarks	Risks	Critical plant substances	Pharmacologically effective dose
Positive monograph, indications: gastrointestinal cramps and inflammatory diseases of the gastrointestinal tract, **standard marketing authorisation:** 7999.99.99, **WHO monograph:** symptomatic treatment of digestive ailments such as dyspepsia, epigastric bloating, impaired digestion, and flatulence, infusions of camomile flowers have been used in the treatment of restlessness and in mild cases of insomnia due to nervous disorders, **ESCOP monograph:** symptomatic treatment of gastrointestinal complaints such as minor spasms, epigastric distension, flatulence and belching, **HMPC monograph:** in progress			**Commission E:** 1 tbsp. (approx. 3 g), 3–4 times a day **ESCOP:** 3 g drug, 3–4 times a day, special dosage instructions for children **WHO:** 2–8 g, 3 times a day, special dosage instructions for children, no administration to children under the age of 3
Positive monograph: for external use, the essential oil is used mostly, which may induce contact allergies when used externally (s. BfR statement: http://www.bfr.bund.de/cm/206/verwendung_von_unverduenntem_teebaumoel_als_kosmetisches_mittel.pdf), **WHO monograph and ESCOP monograph (oil):** for external use		Terpenes (e.g. eucalyptol [=1,8-cineol])	**ESCOP, WHO, Commission E:** for external use
Positive monograph, indications: chronic venous insufficiency (especially pain and feeling of heaviness in legs, night leg cramps, itching, swelling), adjuvant treatment of thrombophlebitis, postthrombotic syndrome, haemorrhoids, lymphatic blockage, **ESCOP monograph:** symptomatic treatment of problems related to varicose veins, such as painful and heavy legs, nocturnal cramps in the legs, itching and swelling, **HMPC monograph:** traditional use, BfArM graduated plan of 23 May 1997 concerning coumarin-containing phytopharmaceuticals and homeopathic drugs up to potency D2 containing sweet clover (Meliloti herba) or sweet woodruff (Gallii odorati herba) (for internal use)	Nausea, vomiting, headache (Roth/Daunderer), contains coumarin, s. BgVV statement: http://www.bfr.bund.de/cm/208/gehalte_an_cumarin_safrol_methyleugenol_und_estragol_in_lebensmitteln.pdf+O245	Coumarin	**Commission E, ESCOP:** equivalent to 3–30 mg coumarin per day, equivalent preparations
Positive monograph, indications: nervous sleep disturbances, functional gastrointestinal disorders, **standard marketing authorisation:** 1149.99.99, **WHO monograph, HMPC monograph:** traditional use, **WHO monograph:** external use, **ESCOP monograph:** tenseness, restlessness and irritability, symptomatic treatment of digestive disorders such as minor spasms			**Commission E, WHO:** 1.5–4.5 g drug, several times a day **ESCOP:** 2–3 g drug, 2–3 times a day, equivalent preparations
		Menthofuran, eucalyptol (=1,8-cineol)	

Stock plant (Latin)	Plant part	F	NF	MP	trad. MP	List A	List B	List C
Mentha pulegium L., syn. Mentha gibraltarica WILLD., Mentha hirtiflora OPIZ ex TOPIZ, Pulegium erectum MILL.	Herb	× S					×	
Mentha spicata L. emend. L. var. crispa BENTH.	Leaf	× T, S						
Mentha suaveolens EHRH.	Herb	× T, S						
Mentha x piperita L., syn. Mentha piperita (L.) HUDS.	Leaf	× T, S		×	×		×	
Menyanthes trifoliata L., syn. Menyanthes palustris TOURN, Trifolium castoris THAL	Leaf	× T		×	×		×	
Momordica charantia L.	Fruit	×		×			×	
Monarda didyma L., syn. Monarda purpurea LAM., Monarda coccinea MICHX.	Flower	× T						
Morus alba L., syn. Morus bombycis KOIDZ., Morus constantinopolitana POIR., Morus indica L.	Leaf		Not NFS					×
Morus alba L., syn. Morus bombycis KOIDZ., Morus constantinopolitana POIR., Morus indica L.	Fruit	×						
Morus nigra L.	Leaf		Not NFS					×
Morus nigra L.	Fruit	×						
Mucuna pruriens (L.) DC., syn. Dolichos pruriens L.	all plant parts					×		
Murraya koenigii (L.) SPRENG.	Leaf	× S						
Musa acuminata X Musa balbisiana, syn. Musa paradisiaca L.	Fruit	×						
Myrciaria dubia (KUNTH) McVAUGH, syn. Psidium dubium KUNTH	Fruit		Not NFS					×
Myristica fragrans HOUTT., syn. Myristica americana ROTTB., Myristica aromatica SWARTZ, Myristica officinalis L.	Dried aril	× S					×	

Remarks	Risks	Critical plant substances	Pharmacologically effective dose
slightly toxic to toxic (Roth/Daunderer), toxic II (Wink/Wyk) contains pulegone (dose restriction acc. to Reg. (EC) No. 1334/2008)	Ingestion of higher doses causes retching, vomiting, increase in blood pressure, central anaesthesia-like paralysis, death by central apnoea, cases of poisoning after abuse as aborticide known (Roth/Daunderer)	Monoterpenes (e.g. pulegone)	
		L-carvone, limonene, eucalyptol (=1,8-cineol)	
Positive monograph, indications: cramps in the gastrointestinal tract, the gall bladder and the biliary tract, standard marketing authorisation: 1499.99.99, WHO monograph (leaves, oil), HMPC monograph: traditional use, ESCOP monograph: symptomatic treatment of digestive disorders, such as flatulence, irritable bowel syndrome, symptomatic treatment of coughs and colds		Menthol, menthone, eucalyptol (=1,8-cineol)	Commission E: 3–6 g drug/day, equivalent preparations ESCOP: 1.5–3 g drug in 150 ml water 3 times a day WHO: 1–3 g drug 3 times a day
Positive monograph, indications: lack of appetite, digestive problems, hardly toxic (+) (Roth/Daunderer)	unpleasant headache, ingestion of high doses causes vomiting and diarrhoea (Roth/Daunderer)		Commission E: 1.5–3 g drug/day, equivalent preparations
Seeds and pericarp of ripe fruits very toxic Ib (Wink/Wyk), therefore only young fruits should be used in food, WHO monograph: emetic, laxative and tonic	Seeds and pericarp of ripe fruits cause severe vomiting and diarrhoea, inhibition of ribosomal biosynthesis (Wink/Wyk)	Cucurbitacins	WHO: 2–15 g drug/day
		Carvacrol, thymol	
toxic + (Roth/Daunderer), toxic II (Wink/Wyk)	hallucinogenic, psychotropic (Wink/Wyk)	Indole alkaloids, levodopa, serotonin	
toxic when abused + (Roth/Daunderer)	Ingestion of high doses causes headache, dizziness, changes in consciousness up to hallucinations, ventricular tachycardia, vertigo, mydriasis, vomiting, abortifacient effect. Safrole contained in the essential oil has a mutagenic effect and demonstrated a carcinogenic effect in animal experiments (Wink/Wyk)	Myristicin, safrole	

Stock plant (Latin)	Plant part	F	NF	MP	trad. MP	List A	List B	List C
Myristica fragrans HOUTT., syn. Myristica americana ROTTB., Myristica aromatica SWARTZ, Myristica officinalis L.	Seed	× S					×	
Nasturtium officinale R. BR., Syn. Baeumerta nasturtium GAERTN., MEY., SCHERB.	Herb	×		×	×		×	
Nelumbo nucifera GAERTN., syn. Nelumbium nelumbo (L.) DRUCE	Flower	× C						
Nepeta cataria L., syn. Nepeta minor MILL., Nepeta vulgaris LAM., Cataria vulgaris MOENCH	Herb	× S, T						
Nerium oleander L., syn. Nerium laurifolium LAM.	Leaf			×		×		
Ocimum basilicum L. syn. Basilicum Indicum et citratum RUMPH., Ocimum album L., Ocimum menthaefolium HOCHST. ex BENTH.	Herb	× S						
Ocimum tenuiflorum L, Ocimum sanctum L., syn. Ocimum album BLANCO, Ocimum virgatum BLANCO	Herb	× S			×			
Olea europaea L., syn. Olea officinarum CRANTZ, Olea pallida SALISB.	Leaf	× T			×			
Ononis spinosa L., syn. Ononis arvensis L., Ononis campestris KOCH et ZIZ.	Root	× T		×			×	

Remarks	Risks	Critical plant substances	Pharmacologically effective dose
Negative monograph: due to risks, toxic when abused + (Roth/Daunderer), toxic II (Wink/Wyk)	Ingestion of high doses causes headache, dizziness, changes in consciousness up to hallucinations, ventricular tachycardia, vertigo, mydriasis, vomiting, abortifacient effect. Safrole contained in the essential oil has a mutagenic effect and demonstrated a carcinogenic effect in animal experiments (Wink/Wyk). Ingestion of approx. 5 g of nutmeg seeds causes a series of mental disorders, which may manifest themselves in slight changes in consciousness up to intense hallucinations. Ingestion of 9 teaspoons of nutmeg powder per day observed to produce an effect similar to that of atropine. The drug has an abortifacient effect in high doses (Commission E monograph)	Myristicin, safrole	
Positive monograph, indications: catarrhs of the respiratory tract, slightly toxic only in larger amounts (+) (Roth/Daunderer)	Contraindications: gastrointestinal complaints observed in rare cases Contraindications: gastrointestinal ulcers, inflammatory kidney diseases (Commission E monograph)	Glucosinolates	**Commission E:** daily dose: 4–6 g drug or 20–30 g fresh herb or 60–150 g freshly pressed juice, equivalent preparations
Negative monograph: due to risks, oleander glycosides prescription-only, oleander pharmacy-only, highly toxic ++ (Roth/Daunderer), extremely toxic Ia (Wink/Wyk)	Numbness of tongue and throat, nausea, vomiting, cramps, disorientation, visual disturbances, diarrhoea, cardiac arrhythmia, apnoea, shock, bradycardia, death may occur 2–5 hours after ingestion (Wink/Wyk, Roth/Daunderer)	cardioactive glycosides	
Negative monograph: due to risks	contains estragole. After metabolic activation, isolated estragole has a mutagenic effect. Animal experiments provide indications of carcinogenic effect. Risks during pregnancy and lactation, for infants and young children as well as during long-term use cannot be ruled out (Schilcher)	Estragole (=methyl chavicol)	
WHO monograph: traditional use			**WHO:** 6–12 g drug/day
Neutral monograph, HMPC monograph: traditional use			
Positive monograph, indications: irrigation therapy of inflammatory diseases of the lower urinary tract, prevention of kidney gravel, **standard marketing authorisation:** 9899.99.99, **ESCOP monograph:** irrigation of the urinary tract, especially in cases of inflammation and renal gravel, and as an adjuvant in treatment of bacterial infections of the urinary tract, **HMPC monograph:** in progress	Contraindications reported (Commission E monograph)		**Commission E:** 6–12 g drug/day, equivalent preparations **ESCOP:** 2–3 g drug, 2 times a day

Stock plant (Latin)	Plant part	F	NF	MP	trad. MP	List A	List B	List C
Origanum majorana L., syn. Amaracus majorana (L.) SCHINZ et THELLUNG, Majorana hortensis MOENCH	Herb	× S						
Origanum majorana L., syn. Amaracus majorana (L.) SCHINZ et THELLUNG, Majorana hortensis MOENCH	Fruit		Not NFS					×
Origanum vulgare L., syn. Origanum anglicum HILL., Origanum creticum L., Origanum latifolium MILL., Origanum venosum WILLD. ex BENTH.	Herb	× S			×			
Orthosiphon aristatus (BLUME) MIQUEL, syn. Orthosiphon stamineus BENTH., Orthosiphon spicatus (THUNB.) BAK., Ocimum grandiflorum BOLD	Leaf	× T		×	×		×	
Oryza sativa L.	Fruit, seed	×						
Osmanthus fragrans (THUN.) LOUR.	Flower		×					
Paeonia officinalis (L.) emend. WILLD., syn. Paeonia feminea MILL., Paeonia foemina GARS.	Flower	× C						
Panax ginseng C.A. MEY., syn. Aralia ginseng (C.A. MEY.) BAILL., Panax quinquefolius L. var. coreensis LIEB.	Root	× T		×			×	
Panax notoginseng CHEN.	Root		Not NFS					×
Panax pseudoginseng WALL.	Root		Not NFS					×
Panax quinquefolius L., syn. Aralia canadensis TOURN., Aureliana canadensis LAFIT., Ginseng quinquefolium WOOD., Panax americanum RAF.	Root		Not NFS	×				×
Panicum miliaceum L., syn. Panicum PERS.	Seed	×						

Remarks	Risks	Critical plant substances	Pharmacologically effective dose
Negative monograph: due to risks	Hydroquinone contained in the drug demonstrates a carcinogenic effect in animal experiments (Commission E monograph)	Hydroquinones and hydro-quinone derivatives (e.g. arbutin)	
Neutral monograph, WHO monograph: traditional use			
Positive monograph, indications: irrigation therapy of bacterial and inflammatory diseases of the lower urinary tract and kidney gravel, **ESCOP monograph:** irrigation of the urinary tract, especially in cases of inflammation and renal gravel, and as an adjuvant in the treatment of bacterial infections of the urinary tract, **HMPC monograph:** traditional use, **standard marketing authorisation:** 1159.99.99	Contraindications: oedemas associated with cardiac and renal insufficiency (Commission E monograph)		**Commission E:** 6 to 12 g drug, equivalent preparations **ESCOP:** 2–3 g in 150 ml water, 2–3 times a day, equivalent preparations
Neutral monograph: slightly toxic (+) (Roth/Daunderer)	Flowers and seeds may induce gastroenteritis accompanied by vomiting, colic and diarrhoea (Roth/Daunderer)		
Positive monograph, indications: tonic for strengthening and invigoration in case of fatigue and faintness, decreasing physical performance and powers of concentration as well as in convalescence, **WHO monograph:** used as a prophylactic and restorative agent for enhancement of mental and physical capacities, in cases of weakness, exhaustion, tiredness, and loss of concentration, and during convalescence, **ESCOP monograph:** decreased mental and physical capacities such as weakness, exhaustion, tiredness and loss of concentration, as well as during convalescence, **HMPC monograph:** in progress, hardly or only slightly toxic (+) (Roth, Daunderer)	Long-term use in higher doses causes diarrhoea, insomnia, euphoria, hypertension and oedemas. "ginseng abuse" syndrome (Roth/Daunderer, Frohne)	Ginsenosides	**Commission E:** 1–2 g drug/day, equivalent preparations **WHO:** 0.5–2 g/day, equivalent preparations, no administration to children
WHO monograph: traditional use			**WHO:** 3–9 g drug/day

Stock plant (Latin)	Plant part	F	NF	MP	trad. MP	List A	List B	List C
Papaver rhoeas L.	Petal		Not NFS					×
Papaver somniferum L., syn. Papaver amplexicaule STOKES, Papaver nigrum CRANTZ, Papaver officinale GMD., Papaver opiiferum FORSK.	Dried latex			×		×		
Papaver somniferum L., syn. Papaver amplexicaule STOKES, Papaver nigrum CRANTZ, Papaver officinale GMD., Papaver opiiferum FORSK.	Seed	×						
Passiflora edulis (AUBL.) SCHUM.	Fruit	×						
Passiflora incarnata L., syn. Granadilla incarnata MEDIK.	Herb	× T		×	×		×	
Passiflora incarnata L., syn. Granadilla incarnata MEDIK.	Flower	× C						
Pastinaca sativa L., syn. Pastinaca opaca BERNH., Pastinaca pratensis H. MART., Pastinaca sativa var. edulis DC., Pastinaca sativa var. longa ALEF., Selinum pastinacae CRANTZ	Root	×						
Paullinia cupana KUNTH ex H.B.K., syn. Paullinia cupana H.B.K., Paullinia sorbilis (L.) MART.	Seed	×		×			×	

Remarks	Risks	Critical plant substances	Pharmacologically effective dose
Neutral monograph			
the clotted milky sap (=dried latex) of plants belonging to the Papaver somniferum species referred to as opium is subject to the provisions of the Narcotic Drugs Act (BtMG) highly toxic ++ (Roth/Daunderer), very toxic Ib (Wink/Wyk)	reduced mental performance, sopor, irregular respiration, anaesthetic muscular relaxation, nausea, vomiting, bradycardia, dizziness, vertigo, tonic-clonic cramps, central paralysis, pulmonary oedema, cerebral oedema, anaphylaxis, acute bilateral loss of hearing, colonic inertia, death by apnoea (Hager, Roth/Daunderer)	Opium alkaloids (e.g. morphine, codeine, papaverine)	
s. BfR statement on opium poppy: http://www.bfr.bund.de/cd/6279, http://www.bfr.bund.de/cd/7409			
Positive monograph, indications: nervous agitation, **standard marketing authorisation:** 1619.99.99, **WHO monograph:** mild sedative for nervous restlessness, insomnia and anxiety. Treatment of gastrointestinal disorders of nervous origin, **ESCOP monograph:** tenseness, restlessness and irritability with difficulty in falling asleep, **HMPC monograph:** traditional use; information provided in the literature stating that passion flower contains harmala alkaloids not confirmed in more recent studies. Only traces of harmala alkaloids (below 1 ppm) detected in individual samples. Whether these amounts are sufficient for CNS activity needs to be investigated in in-vivo studies. (Hänsel/Sticher, Hager)	HCN release due to cyanogenic glycosides (Passiflora incarnata contains at most traces of these compounds (Hager)), release of harmala alkaloids is likely to cause hallucinogenic effects (Wink/Wyk)	cyanogenic glycosides, harmala alkaloids	**Commission E:** 4–8 g drug/day, equivalent preparations **ESCOP, WHO:** 0.5–2 g drug, 3–4 times a day
Information provided in the literature stating that passion flower contains harmala alkaloids not confirmed in more recent studies. Only traces of harmala alkaloids (below 1 ppm) detected in individual samples. Whether these amounts are sufficient for CNS activity needs to be investigated in in-vivo studies. (Hänsel/Sticher, Hager)	Release of harmala alkaloids is likely to cause hallucinogenic effects (Wink/Wyk)	Harmala alkaloids	
		Myristicin, furocoumarin	
mildly toxic III (Wink/Wyk), toxic only when abused + (Roth/Daunderer), **ESCOP monograph:** tonic in cases of fatigue, exhaustion or convalescence, **HMPC monograph:** traditional use	CNS stimulant, ingestion of higher doses causes poisoning accompanied by headache, vertigo, cardiac palpitation, sleep disturbances, agitation, vomiting, diarrhoea, severe diuresis (Wink/Wyk)	Caffeine, saponins	**ESCOP:** 1–3 g drug/day, equivalent preparations

Stock plant (Latin)	Plant part	F	NF	MP	trad. MP	List A	List B	List C
Pausinystalia yohimbe (K. SCHUM.) PIERRE ex BEILLE, syn. Corynanthe yohimbe K. SCHUM.	all plant parts			×		×		
Peganum harmala L.	all plant parts			×		×		
Petasites spp., e.g. Petasites hybridus (L.) GAERTN., MEY. et SCHERB.	all plant parts			×		×		
Petroselinum crispum (MILL.) NYM. ex A.W. HILL., syn. Apium hortense E.H.L. KRAUSE, Apium petroselinum L., Apium vulgare LAM., Carum petroselinum BENTH. et HOOK.	Seed	× T					×	
Petroselinum crispum (MILL.) NYM. ex A.W. HILL., syn. Apium hortense E.H.L. KRAUSE, Apium petroselinum L., Apium vulgare LAM., Carum petroselinum BENTH. et HOOK.	Herb	× S		×			×	
Petroselinum crispum (MILL.) NYM. ex A.W. HILL., syn. Apium hortense E.H.L. KRAUSE, Apium petroselinum L., Apium vulgare LAM., Carum petroselinum BENTH. et HOOK.	Root	×		×			×	

Remarks	Risks	Critical plant substances	Pharmacologically effective dose
Negative monograph: due to risks associated with Yohimbe bark, Yohimbe pharmacy-only, yohimbic acid and its esters prescription-only, very toxic Ib (Wink/Wyk), highly toxic ++ (Roth/Daunderer) s. BfR statement: http://www.bfr.bund.de/cm/350/risikobewertung-von-pflanzen-und-pflanzlichen-zubereitungen.pdf	Vertigo, hypersalivation, faintness, cold spills, sweats, nervous excitement, cardiac palpitation, insomnia, gastric complaints, headache, gastrointestinal complaints, increase in blood pressure and tachycardia, tremor, irritability (Hager)	Alkaloids (e.g. yohimbine)	
WHO monograph: treatment of different forms of myasthenia, myopathy and atony of the bowels, highly toxic ++ (Roth/Daunderer), very toxic Ib (Wink/Wyk)	psychotropic, hallucinogenic, hypersalivation, vomiting, muscular tremor (Hager, Wink/Wyk, Roth/Daunderer)	Indole alkaloids (e.g. harmine, harmaline), quinazoline alkaloids	**WHO monograph:** a teaspoon of the preparation of a teaspoon of drug in one cup of water, 3–4 times a day
Positive monograph: for Petasites hybridus rhizoma (butterbur rootstock), **indications:** adjuvant treatment of acute cramps in the lower urinary tract, **Negative monograph:** due to risks associated with Petasites hybridus folium (butterbur leaves), classification of Petasitidis folium as substance of concern (AMK), BfArM graduated plan of 5 June 1992 concerning medicinal products containing pyrrolizidine alkaloids, toxic + (Roth/Daunderer), mildly toxic II to toxic III (Wink/Wyk)	All plant parts of butterbur contain varying amounts of hepatotoxic pyrrolizidine alkaloids, which demonstrated a mutagenic and carcinogenic effect in animal experiments (Schilcher)	Pyrrolizidine alkaloids	**Commission E:** 4.5–7 g drug/day
Negative monograph: due to risks, Roth/Daunderer: hardly toxic (+), mildly toxic III (Wink/Wyk)	In high doses, the essential oil in parsley seeds and the phenyl propanoid derivative apiol contained in it induce vascular congestions and enhance the contractility of the smooth muscles of the bladder, the intestines and especially the uterus. Parsley seeds and parsley oil therefore formerly used as aborticide; irritation, damage to renal epithelial tissues and cardiac arrhythmia reported, higher doses of apiol in parsley may cause fatty liver disease, emaciation, bloody stool, profuse haemorrhage of mucous membranes and haemorrhagic-inflammatory infiltrations in the gastrointestinal tract, haemolysis, methaemoglobinuria and anuria (Commission E monograph). Toxic effects are not expected in usual doses (Roth/Daunderer), phototoxic	Furocoumarins, phenyl propanoids (e.g. myristicin, apiol)	
Positive monograph, indications: irrigation therapy of diseases of the lower urinary tract and kidney gravel, Roth/Daunderer: hardly toxic (+), mildly toxic III (Wink/Wyk)	like seeds, but lower essential oil content, phototoxic	Furocoumarins, phenyl propanoids (e.g. myristicin, apiol)	**Commission E:** 6 g drug/day, equivalent preparations
Positive monograph, indications: irrigation therapy of diseases of the lower urinary tract and kidney gravel, Roth/Daunderer: hardly toxic (+), mildly toxic III (Wink/Wyk)	like seeds, but lower essential oil content, phototoxic	Furocoumarins, phenyl propanoids (e.g. myristicin, apiol)	**Commission E:** 6 g drug/day, equivalent preparations

Stock plant (Latin)	Plant part	F	NF	MP	trad. MP	List A	List B	List C
Peumus boldus MOL., syn. Peumus fragrans PERS., Boldea fragrans JUSS., Boldea fragrans C. GAY	Leaf			×	×	×		
Pfaffia paniculata (MART.) KUNTZE	Root		Not NFS					×
Phaseolus vulgaris L., syn. Phaseolus nanus L.	Pod	× b		×			×	
Phaseolus vulgaris L., syn. Phaseolus nanus L.	Seed	× b						
Phoenix dactylifera L.	Fruit	×						
Phyllanthus emblica L., Emblica officinalis GAERTN., Phyllanthus mairei LÉV.	Fruit		Not NFS					×
Physostigma venenosum BALFOUR	Seed			×		×		
Phytolacca americana L., syn. Phytolacca decandra L., Phytolacca vulgaris CRANTZ.	Fruit					×		
Picea abies (L.) KARSTEN, syn. Abies excelsa LAM. et DC., Abies picea MILLER, Pinus abies L.	Shoot (tip)	× T		×			×	

Remarks	Risks	Critical plant substances	Pharmacologically effective dose
Positive monograph, indications: mild gastrointestinal cramps, digestive problems, **standard marketing authorisation:** 2329.99.99, **ESCOP monograph:** minor hepatobiliary dysfunction, symptomatic treatment of mild digestive disturbances, **HMPC monograph:** traditional use, toxic II (Wink/Wyk), toxic + (Roth/Daunderer)	neurotoxic effect observed, ingestion of higher doses causes paralysis, hallucinations and CNS disorders (Wink/Wyk), contraindications: occlusion of biliary tract, severe liver diseases, pregnancy (because of ascaridole content). In case of gallstones, use only after consultation with a physician. (Commission E monograph)	Alkaloids (e.g. boldine), monoterpenes (e.g. ascaridole)	**Commission E:** average daily dose: 3.0 g drug, equivalent preparations **ESCOP:** 2–5 g drug/day
Positive monograph, indications: slightly diuretic, **standard marketing authorisation:** 8499.99.99; in unprocessed form mildly toxic II to toxic III (Wink/Wyk), highly toxic ++ (Roth/Daunderer), heat thoroughly before consumption, **HMPC monograph:** traditional use	Seeds and pods contain lectin ("phasin"). Since phasin is a protein, it is destroyed by cooking, but not by drying. Symptoms usually occur after 2 to 3 hours. The severity of the disease varies from case to case. Symptoms include nausea, abdominal pain and vomiting, gastroenteritis and cramps. May be accompanied by bloody diarrhoea. Besides fever, shivering and sweats, convulsive fits and shock may occur. (Bonn Poison Control Centre)	Lectin ("phasin")	**Commission E:** 5–15 g drug/day, equivalent preparations
in unprocessed form mildly toxic II to toxic III, highly toxic ++ (Roth/Daunderer), heat thoroughly before consumption	Seeds and pods contain lectin ("phasin"). Since phasin is a protein, it is destroyed by cooking, but not by drying. Symptoms usually occur after 2 to 3 hours. The severity of the disease varies from case to case. Symptoms include nausea, abdominal pain and vomiting, gastroenteritis and cramps. May be accompanied by bloody diarrhoea. Besides fever, shivering and sweats, convulsive fits and shock may occur. (Bonn Poison Control Centre)	Lectin ("phasin")	
poisonous plant, extremely toxic (Roth/Daunderer), extremely toxic (Wink/Wyk), physostigmine prescription-only, Physostigma species pharmacy-only	Paraesthesia, inflammation of mucous membranes, tremor, tachycardia, persistent vertigo, death by cardioplegia and respiratory arrest (Wink/Wyk)	Alkaloids (e.g. physostigmine)	
toxic + (Roth/Daunderer), toxic II (Wink/Wyk)	Nausea, vomiting, diarrhoea, abdominal pain, colic, headache, circulatory distress, abortion may occur, fatal poisoning possible (Wink/Wyk, Roth/Daunderer)	Lectins, triterpene saponins	
Positive monograph, indications: catarrhs of the respiratory tract			**Commission E:** 5–6 g drug/day

Stock plant (Latin)	Plant part	F	NF	MP	trad. MP	List A	List B	List C
Pilocarpus jaborandi HOLMES, syn. Pilocarpus cearensis RIZZINI, Pilocarpus officinalis POEHL	Leaf			×		×		
Pimenta officinalis LINDL., syn. Pimenta officinalis BERG, Pimenta vulgaris LINDL., Eugenia pimenta DC., Myrtus pimenta L., Pimenta dioica (L.) MERR.	Fruit	× S						
Pimpinella anisum L., syn. Anisum vulgare GAERTN.	Fruit	× T, S		×	×		×	
Pimpinella major (L.) HUDS., syn. Apium pimpinella CARUEL	Herb	× S						
Pimpinella major (L.) HUDS., syn. Apium pimpinella CARUEL	Root	× T,S		×	×		×	
Pimpinella saxifraga L., syn. Pimpinella alpestris SPRENG.	Herb	× S						
Pimpinella saxifraga L., syn. Pimpinella alpestris SPRENG.	Root	× T,S		×	×		×	
Pinus mugo TURRA, syn. Pinus echinata HORT., Pinus montana MILLER	Shoot (tip)	× T						
Pinus sylvestris L., syn. Pinus nigra MILL.	Shoot (tip)	× T		×			×	
Piper cubeba L.f., syn. Cubeba officinalis RAF.	Fruit	× S					×	

Remarks	Risks	Critical plant substances	Pharmacologically effective dose
Pilocarpine prescription-only, Pilocarpus species pharmacy-only, extremely toxic (Roth/Daunderer), very toxic (Wink/Wyk), substance of concern (AMK)	parasympathomimetic effect, hypersalivation, sweating, visual disturbances, vomiting, diarrhoea, uterus contractions, abortion, bradycardia, cardiac palpitation, convulsions, tremor, dyspnoea, arrhythmia, vesical tenesmus, headache, vertigo, death by cardiac arrest, leucocytosis (Wink/Wyk, Roth/Daunderer)	Imidazole alkaloids (e.g. pilocarpine)	
Positive monograph, indications: digestive problems, catarrhs of the respiratory tract, **standard marketing authorisation:** 8099.99.99, **WHO monograph (fruit and oil):** treatment of dyspepsia and mild inflammation of the respiratory tract, **ESCOP monograph:** dyspeptic complaints such as mild spasmodic gastrointestinal complaints, bloating, flatulence, catarrh of the upper respiratory tract, **HMPC monograph:** traditional use	allergic reactions of skin, respiratory tract and gastrointestinal tract occur occasionally (Commission E monograph)	Trans-anethole, furocoumarins, estragole (=methyl chavicol)	**Commission E, ESCOP, WHO:** 3 g drug/day, equivalent preparations
Neutral monograph	phototoxic (Hager)	Furocoumarins	
Positive monograph, indications: catarrhs of the upper respiratory tract	phototoxic (Hager)	Furocoumarins	**Commission E:** 6–12 g drug/day, equivalent preparations
Neutral monograph	phototoxic (Hager)	Furocoumarins	
Positive monograph, indications: catarrhs of the upper respiratory tract	phototoxic (Hager)	Furocoumarins	**Commission E:** 6–12 g drug/day, equivalent preparations
Positive monograph, indications: catarrhs of the upper respiratory tract			**Commission E:** 2–3 g drug/day, equivalent preparations
toxic II (Wink/Wyk), toxic + (Roth/Daunderer)	Ingestion of higher doses causes painful urinary tract inflammations, pain in kidneys and bladder, albuminuria, tachycardia, nausea, vomiting, diarrhoea, CNS disorders, spasmodic urinary retention, accelerated pulse, cardialgia, diarrhoea and skin rash (Wink/Wyk, Roth/Daunderer)	Cubebin, cubebic acid	

Stock plant (Latin)	Plant part	F	NF	MP	trad. MP	List A	List B	List C
Piper methysticum G. FORST., syn. Macropiper latifolium MIQ.	Rhizome			×		×		
Piper nigrum L., syn. Muldera multinervis MIQ., Piper aromaticum LAM., Piper trioicum ROXB.	Fruit	× S						
Piscidia erythrina L., syn. Comptosema pinnatum BENTH., Piscidia piscipula (L.) SARG.	Root cortex					×		
Pistacia vera L., syn. Pistacia narbonensis L., Pistacia nigricans CRANTZ, Pistacia officinarum AITON	Seed	×						
Pisum sativum L.	Seed	×						
Plantago lanceolata L., Arnoglossum lanceolatum (L.) GRAY, Plantago flexuosa GAUD ex RAPIN, Plantago lanceofolia SALISB.	Herb	× T		×	×		×	
Plantago major L., syn. Plantago asiata L., Plantago latifolia SALISB.	Herb		Not NFS	×			×	

Remarks	Risks	Critical plant substances	Pharmacologically effective dose
withdrawal of marketing authorisation due to risks following the graduated plan procedure (BfArM notification of 21 Dec 2007: http://www.bfarm.de/DE/Pharmakovigilanz/risikoinfo/stufenplanverf/Liste/stp-kavakava.html; jsessionid=7AF6D40DED621345B74BFDC9AD5 1_cid332), **Positive monograph, indications:** states of nervous anxiety, stress and agitation, **WHO monograph:** short-term symptomatic treatment of mild states of anxiety or insomnia, due to nervousness, stress or tension, **ESCOP monograph:** anxiety, tension and restlessness arising from various causes of non-psychotic origin, substance of concern (AMK): kava-kava and kavain, toxic + (Roth/Daunderer), mildly toxic (Wink/Wyk)	severe hepatotoxic reactions (Wink/Wyk, Roth/Daunderer, BfArM notification on graduated plan)	Kavapyrones (e.g. kavain)	**ESCOP:** equivalent to 60–120 mg kavalactones/day, (no data available for children) **WHO:** 60–210 mg kavapyrones/day
Black pepper: dried, unripe fruit White pepper: ripe fruit without fruit pulp Green pepper: air-dried, freeze-dried, deep-frozen or pickled unripe fruit Pink pepper: fully ripe fruit (Teuscher)			
toxic + (Roth/Daunderer), very toxic Ib (Wink/Wyk)	Vomiting, hypersalivation, sweats, dizziness and tremor, first enhanced then depressed respiration, central paralysis, cardioplegia, used as fish and arrow poison (Wink/Wyk)	Isoflavones (jamaicin, munetone, ichthynone, rotenone)	
Positive monograph, indications: catarrhs of the respiratory tract, inflammatory changes of the oral and pharyngeal mucosa, **standard marketing authorisation:** 1289.99.99, **ESCOP monograph:** catarrhs of the respiratory tract, temporary, mild inflammations of the oral and pharyngeal mucosa, **HMPC monograph:** traditional use			**ESCOP, Commission E:** 3–6 g drug/day, equivalent preparations **WHO monograph:** 2–4 g drug, 3 times a day
WHO monograph, indications: catarrh of the respiratory tract, cough, bronchitis, inflammatory alterations of the oral mucosa, phlegm congestion, nephritis, cystitis, urinary retention, dysentery, epistaxis (nosebleed) and excessive diaphoresis (49, 50). It is also used for the treatment of diarrhoea and constipation			

Stock plant (Latin)	Plant part	F	NF	MP	trad. MP	List A	List B	List C
Plantago ovata FORSSKAL, syn. Plantago isphagula ROXB. ex FLEMMING, Plantago brunnea MORRIS	Seed husk	×		×			×	

Remarks	Risks	Critical plant substances	Pharmacologically effective dose
higher swelling number than psyllium seeds, risk of intestinal occlusion after overdose, Positive monograph, indications: habitual constipation, conditions in which easy defaecation is desirable, e.g. anal fissures, haemorrhoids, after rectal-anal surgical procedures and during pregnancy, adjuvant treatment of diarrhoea of various origins as well as irritable bowel, **WHO monograph:** a bulk-forming laxative used therapeutically for restoring and maintaining bowel regularity, treatment of chronic constipation, temporary constipation due to illness or pregnancy, irritable bowel syndrome and constipation related to duodenal ulcer or diverticulitis, also indicated for stool softening in the case of haemorrhoids, or after anorectal surgery, short-term use for the symptomatic treatment of diarrhoea of various etiologies, **ESCOP monograph:** treatment of occasional constipation, conditions in which easy defication with soft stools desireable, e.g. in cases of anal fissures or haemorrhoids, after rectal or anal surgery, and during pregnancy; conditions which need an increased daily intake of fibre, e.g. irritable bowel syndrome, adjuvant symptomatic therapy in cases of diarrhoea from various causes, as an adjunct to a low fat diet in the treatment of mild to moderate hypercholesterolaemia, **HMPC monograph:** habitual constipation; in conditions in which easy defaecation with soft stool is desirable, e.g. in cases of painful defaecation after rectal or anal surgery, anal fissures and haemorrhoids; in patients to whom an increased daily fibre intake may be advisable e.g. as an adjuvant in constipation predominant irritable bowel syndrome, as adjuvant to diet in hypercholesterolemia	Side effects: hypersensitivity reactions may occur in individual cases, contraindications: pathologic stenosis of gastrointestinal tract, imminent or existing intestinal occlusion (ileus), difficult-to-control diabetes mellitus (Commission E monograph), oesophageal occlusion or obstruction up to intestinal occlusion may occur on improper use (HMPC monograph)		**Commission E, ESCOP:** 4–20 g drug/day, equivalent preparations (children!) **HMPC:** 7–20 g drug/day, equivalent preparations, special doses for children

Stock plant (Latin)	Plant part	F	NF	MP	trad. MP	List A	List B	List C
Plantago ovata FORSSKAL, syn.Plantago isphagula ROXB. ex FLEMMING, Plantago brunnea MORRIS	Seed	×		×			×	
Plantago psyllium L., syn. Plantago afra L., Plantago (L.) MIRB.	Seed	×		×			×	
Podophyllum peltatum L.	Root, resin			×		×		

Remarks	Risks	Critical plant substances	Pharmacologically effective dose
Positive monograph, indications: habitual constipation, conditions in which easy defaecation is desirable, e.g. anal fissures, haemorrhoids, after rectal-anal surgical procedures and during pregnancy, adjuvant treatment of diarrhoea of various origins as well as irritable bowel, **standard marketing authorisation:** 1549.99.99, **WHO monograph:** as a bulk-forming laxative used to restore and maintain regularity, Semen Plantaginis is indicated in the treatment of chronic constipation, temporary constipation due to illness or pregnancy, irritable bowel syndrome, constipation related to duodenal ulcer or diverticulitis. It is also used to soften the stools of those with haemorrhoids, or after anorectal surgery, traditional use, **ESCOP monograph:** treatment of occasional constipation, conditions in which easy defication with soft stools desireable, e.g. in cases of anal fissures or haemorrhoids, after rectal or anal surgery, and during pregnancy; conditions which need an increased daily intake of fibre, e.g. irritable bowel syndrome, adjuvant symptomatic therapy in cases of diarrhoea from various causes, **HMPC monograph:** treatment of habitual constipation. In conditions in which easy defaecation with soft stool is desirable, e.g. in cases of painful defaecation after rectal or anal surgery, anal fissures and haemorrhoids.	allergic reactions observed in rare cases, especially with comminuted drug and liquid preparations (Commission E monograph), oesophageal occlusion or obstruction up to intestinal occlusion may occur on improper use (HMPC monograph)		**Commission E:** 12–40 g drug/day, equivalent preparations **ESCOP:** 7–30 g/day, equivalent preparations, special doses for children and for the indication of hypercholesterolaemia **WHO:** 7.5 g drug in 240 ml water, 1–3 times a day, special doses for children **HMPC:** 8–40 g drug/day, equivalent preparations, special doses for children
Positive monograph, indications: habitual constipation, irritable colon, **standard marketing authorisation:** 1509.99.99, **WHO monograph:** as a bulk-forming laxative used to restore and maintain regularity, Semen Plantaginis is indicated in the treatment of chronic constipation, temporary constipation due to illness or pregnancy, irritable bowel syndrome, constipation related to duodenal ulcer or diverticulitis, it is also used to soften the stools of those with haemorrhoids, or after anorectal surgery, **ESCOP monograph:** treatment of occasional constipation, conditions in which easy defication with soft stools desireable, e.g. in cases of anal fissures or haemorrhoids, after rectal or anal surgery, and during pregnancy, adjuvant symptomatic therapy in cases of diarrhoea from various causes, **HMPC monograph:** treatment of habitual constipation, in conditions in which easy defecation with soft stool is desirable, e.g. in cases of painful defecation after rectal or anal surgery, anal fissures and haemorrhoids.	Contraindications: oesophageal and gastrointestinal stenosis, side effects: allergic reactions observed in rare cases, especially with comminuted drug (Commission E monograph), oesophageal occlusion or obstruction up to intestinal occlusion may occur on improper use (HMPC monograph)		**Commission E, ESCOP:** 10–30 g drug/day, equivalent preparations (special doses for children) **WHO:** 7.5 g drug in 240 ml water, 1–3 times a day, special doses for children **HMPC:** 25–40 g drug/day, equivalent preparations, special doses for children
Positive monograph for external use, also toxic when used externally, root and preparations thereof as well as various substances of P. peltatum prescription-only, use as vermifuge, laxative and diuretic reported (Roth/Daunderer), very toxic Ib (Wink/Wyk), toxic + to highly toxic ++ (Roth/Daunderer)	Plant cytostatic, strong mitotic poison, haemorrhagic gastroenteritis, bloody-watery diarrhoea, impaired coordination, cramps, tachycardiac arrhythmia, paraesthesia, nephritis, coma, death by apnoea (Roth/Daunderer)	Podophyllin, lignans (e.g. podophyllotoxin, peltatin)	

Stock plant (Latin)	Plant part	F	NF	MP	trad. MP	List A	List B	List C
Pogostemon patchouly PELLET., syn. Pogostemon cablin (BLANCO) BENTH., Pogostemon intermedium BENTH.	Leaf		Not NFS					×
Polygala senega L.	Root	× T		×	×		×	
Polygonatum multiflorum (L.) ALL., syn Convallaria multiflora L.	all plant parts					×		
Polygonum aviculare L., syn. Polygonum heterophyllum LINDM.	Herb	× T		×			×	
Potentilla anserina L., syn. Argentina anserina RYDBERG, Argentina vulgaris LAM., Dactylophyllum anserina SPENNER	Herb	× T		×			×	
Potentilla erecta (L.) RÄUSCHEL, syn. Potentilla tormentilla NECKER, Potentilla tormentilla SCHRANK, Potentilla tormentilla STOKES, Tormentilla erecta L., Tormentilla officinalis CURT.	Rhizome	× F		×	×		×	
Primula elatior (L.) HILL, Primula veris L.	Flower	× C		×	×		×	
Primula veris L., syn. Primula coronaria SALISB. Primula montana REUT., Primula odorata GILIB. and Primula elatior (L.) HILL	Root	× T		×	×		×	
Prunus amydalus var. sativa (L.) FOCKE, syn. Amygdalus communis L. var. dulcis DC.	Flower		Not NFS					×
Prunus amydalus var. sativa (L.) FOCKE, syn. Amygdalus communis L. var. dulcis DC.	Seed	×						
Prunus amygdalus BATSCH var. amara (DC.) FOCKE	Seed	× F					×	
Prunus armeniaca L., syn. Armeniaca vulgaris LAM.	Fruit	×						

Remarks	Risks	Critical plant substances	Pharmacologically effective dose
Positive monograph, indications: catarrhs of the upper respiratory tract, **ESCOP monograph:** productive cough, catarrh of the respiratory tract, chronic bronchitis, **WHO monograph:** traditional use	contains saponins, long-term use causes gastrointestinal irritation (Commission E monograph)		**Commission E, ESCOP, WHO:** 1.5–3 g drug/day, equivalent preparations, administration to children only under medical supervision
toxic + (Roth/Daunderer), mildly toxic II to toxic III (Wink/Wyk)	Nausea, diarrhoea, vomiting, used as emetic in traditional Russian medicine (Roth/Daunderer)	steroidal saponins	
Positive monograph, indications: mild catarrhs of the respiratory tract, inflammatory changes of the oral and pharyngeal mucosa, **WHO monograph:** supportive treatment of gingivitis, used for the relief of cough and cold symptoms			**Commission E, WHO monograph:** 4–6 g drug/day, equivalent preparations
Positive monograph, indications: mild dysmenorrhoea, adjuvant treatment of mild, non-specific, acute forms of diarrhoea, mild inflammations of the oral and pharyngeal mucosa, **standard marketing authorisation:** 9599.99.99	Symptoms of irritable stomach may be aggravated (Commission E monograph)		**Commission E:** 4–6 g drug/day, equivalent preparations
Positive monograph, indications: non-specific, acute forms of diarrhoea, mild inflammations of the oral and pharyngeal mucosa, **HMPC monograph:** traditional use, s. BfR statement: http://www.bfr.bund.de/cm/ 350/risikobewertung-von-pflanzen-und-pflanzlichen-zubereitungen.pdf	Sensitive patients may experience gastric complaints (Commission E monograph)		**Commission E:** 4–6 g drug/day, equivalent preparations
Positive monograph, indications: catarrhs of the respiratory tract, **WHO monograph:** traditional use, **standard marketing authorisation:** 1659.99.99, **HMPC monograph:** traditional use	gastric complaints and nausea observed in rare cases, contraindications reported (Commission E monograph)		**Commission E:** 2–4 g drug/day, equivalent preparations
Positive monograph, indications: catarrhs of the respiratory tract, **standard marketing authorisation:** 2389.99.99, **ESCOP monograph:** productive cough, catarrh of the respiratory tract, chronic bronchitis, **HMPC monograph:** traditional use, mildly toxic III (Wink/Wyk)	Side effects: gastric complaints and nausea observed in rare cases, contraindications reported (Commission E monograph)	Saponins	**Commission E, ESCOP:** 0.5–1 g (5–10 g) drug/day, equivalent preparations
Guidelines for oils seeds as well as mixtures and sweets produced thereof laid down in the German Food Code			
Guidelines for oils seeds as well as mixtures and sweets produced thereof laid down in the German Food Code: Almonds may contain up to 5 wt% bitter almonds, very toxic Ib (Wink/Wyk), toxic + to highly toxic ++ (Roth/Daunderer)	Hydrocyanic acid (HCN) may be released on ingestion, HCN is a strong respiratory inhibitor, ingestion of higher doses causes burning in the throat, sweating, abdominal pain, vomiting, facial blush, hypersalivation, convulsions, respiratory and cardiac arrest (Wink/Wyk)	cyanogenic glycosides (e.g. amygdalin)	

Stock plant (Latin)	Plant part	F	NF	MP	trad. MP	List A	List B	List C
Prunus armeniaca L., syn. Armeniaca vulgaris LAM.	Seed	×		×			×	
Prunus avium L.	Fruit	×						
Prunus avium L.	Stalk		Not NFS					×
Prunus cerasus L., Cerasus vulgaris MILL., Cerasus caproniana DC., Prunus caproniana GAUD., Druparia cerasus CLAIRV.	Fruit	×						
Prunus cerasus L., Cerasus vulgaris MILL., Cerasus caproniana DC., Prunus caproniana GAUD., Druparia cerasus CLAIRV.	Stalk		Not NFS					×
Prunus domestica L., syn. Prunus communis HUDS. Non ARCANG., Prunus sativa ROUY et CAMUS	Fruit	×						
Prunus persica (L.) SIEB. et ZUCC., syn. Persica vulgaris MILL., Amygdalus persica L.	Fruit	×						
Prunus spinosa L., syn. Druparia spinosa CLAIRV.	Flower	× T						
Prunus spinosa L., syn. Druparia spinosa CLAIRV.	Fruit	×						
Psidium guajava L.	Fruit	×						
Psidium guajava L.	Leaf		Not NFS	×			×	
Psychotria viridis RUIZ et PAV.	all plant parts					×		
Pteridium aquilinum (L.) KUHN, syn. Pteris aquilina L., Asplenium aquilinum (L.) BERNH.	all plant parts					×		
Pterocarpus santalinus L.f., syn. Lingoum santalinum (L.f.) KUNTZE.	Wood	× S, T						
Ptychopetalum olacoides BENTH.	Wood		Not NFS					×
Pueraria lobata (WILLD.) OHWI, syn. Pueraria thunbergiana (SIEB. et ZUCC.) BENTH., Pueraria hirsuta (THUNB.) SCHNEID. non KURZ, Dolichos japonicus HORT.	Root	×						

Remarks	Risks	Critical plant substances	Pharmacologically effective dose
WHO monograph: symptomatic treatment of asthma, cough with profuse expectoration and fever. The seed oil is used for treatment of constipation; Guidelines for oils seeds as well as mixtures and sweets produced thereof: http://www.bmelv.de/SharedDocs/Downloads/Ernaehrung/Lebensmittelbuch/LeitsaetzeOelsamen.pdf?__blob=publicationFile	Bitter apricot kernels have a high cyanogenic glycoside content, e.g. amygdalin, which breaks down to hydrogen cyanide during digestion, symptoms of acute poisoning include e.g. cramps, vomiting, dyspnoea, culminating in death by apnoea (BfR, FSA)	cyanogenic glycosides	**WHO:** 3–9 g drug/day, no administration to children
Neutral monograph			
Positive monograph: for external use (mouthwash)			**Commission E:** 2–4 g drug/day, equivalent preparations
WHO monograph: acute diarrhoea, gingivitis, rotaviral enteritis			**WHO:** 500 mg drug, 3–4 times a day, no administration to children under the age of 12
mildly toxic III (Wink/Wyk), used together with Banisteriopsis caapi to produce ayahuasca, a narcotic, DMT is subject to the provisions of the BtMG	psychotropic, hallucinogenic (Wink/Wyk)	Tryptamine alkaloids (e.g. N,N-dimethyl-tryptamine=DMT)	
all plant parts very toxic Ib (Wink/Wyk), toxic + (Roth/Daunderer)	Increased incidence of gastric and oesophageal carcinoma reported (Frohne/Pfänder), ptaquiloside contained in the drug is highly mutagenic and causes stomach and bladder cancer, thiaminase contained in the drug destroys vitamin B1 (Wink/Wyk)	Ptaquiloside, thiaminase, cyanogenic glycosides	
Neutral monograph			
Neutral monograph			
		Isoflavones	

Stock plant (Latin)	Plant part	F	NF	MP	trad. MP	List A	List B	List C
Pulsatilla pratensis (L.) MILL., syn. Anemone intermedia SCHULT., Anemona pratensis L., Pulsatilla breynii RUPR.	all plant parts			×		×		
Pulsatilla vulgaris MILL., syn. Anemona pulsatilla L.	all plant parts			×		×		
Punica granatum L., syn. Granatum punicum ST. LAG., Punica florida SALISB.	Fruit	×						
Punica granatum L., syn. Granatum punicum ST. LAG., Punica florida SALISB.	Peel		Not NFS	×			×	
Punica granatum L., syn. Granatum punicum ST. LAG., Punica florida SALISB.	Bark		×	×				
Pyrus communis L.	Fruit	×						
Quassia amara L., syn. Quassia alatifolia STOKES, Quassia cocea VAHL, Quassia gujanensis D. DIETR.	Wood	× F					×	
Quercus robur L., syn. Quercus femina MILL., Quercus fructipendula SCHRANK, Quercus germanica LASCH	Bark	× T		×	×		×	
Quillaja saponaria MOLINA, syn. Quillaja molinae DC.	Bark		Not NFS					×
Raphanus sativus L., syn. Raphanus officinalis CRANTZ, Raphanus raphanistrum var. sativus BECK	Root	×		×			×	
Rauvolfia serpentina (L.) BENTH. ex KURZ, syn. Ophioxylon obversum MIQ., Rauvolfia observa (MIQ.) BAILL., Rauvolfia trifoliata (GAERTN.) BAILL.	Root			×		×		

Remarks	Risks	Critical plant substances	Pharmacologically effective dose
Negative monograph: due to risks, Pulsatillae herba and preparations thereof prescription-only, pasque flower pharmacy-only, toxic + (Roth/Daunderer) toxic Ib to very toxic II (Wink/Wyk)	central paralysis, gastrointestinal complaints (nausea, vomiting, diarrhoea), irritation of mucous membranes, irritation of kidneys and lower urinary tract (Wink/Wyk, Roth/Daunderer), protoanemonin contained in the drug has an abortifacient and teratogenic effect (Schilcher)	Protoanemonin, saponins	
Negative monograph: due to risks, Pulsatillae herba and preparations thereof prescription-only, pasque flower pharmacy-only, toxic + (Roth/Daunderer) toxic Ib to very toxic II (Wink/Wyk)	central paralysis, gastrointestinal complaints (nausea, vomiting, diarrhoea), irritation of mucous membranes, irritation of kidneys and lower urinary tract (Wink/Wyk, Roth/Daunderer), protoanemonin contained in the drug has an abortifacient and teratogenic effect (Schilcher)	Protoanemonin, saponins	
WHO monograph for bark and pericarp: chronic diarrhoea, dysentery, gingivitis and intestinal parasites			**WHO:** pericarp 2.5–4.6 g/day, no administration to children under the age of 12
WHO monograph for bark and pericarp: diarrhoea, intestinal parasites	Ingestion of higher doses (more than 80 g) may cause nausea, diarrhoea, haemorrhage, vertigo, fever, tremor, circulatory collapse, visual disturbances (WHO monograph)		**WHO:** cortex (3–9 g drug/day), no administration to children under the age of 12
mildly toxic II to toxic III (Wink/Wyk), slightly toxic + (Roth/Daunderer), use as flavouring governed by Reg. (EC) No. 1334/2008	Ingestion of high doses causes digestive problems and CNS disorders (Wink/Wyk), gastric irritation, vomiting, contraindicated in pregnancy (Roth/Daunderer)	Quassin	
Positive monograph, indications: external use: inflammatory skin diseases, internal use: non-specific, acute forms of diarrhoea, local treatment of mild inflammations in the mouth and throat as well as in the genital and anal region, **HMPC monograph:** traditional use, **standard marketing authorisation:** 9099.99.99, mildly toxic (Wink/Wyk), non-toxic or hardly toxic (+) (Roth/Daunderer)	Interactions: absorption of alkaloids and other alkaline medicinal products may be reduced or inhibited (Commission E monograph)		**Commission E:** 3 g drug/day, equivalent preparations
may be slightly toxic (Roth/Daunderer)	Overdose causes gastrointestinal complaints accompanied by gastric pain and diarrhoea, irritating to mucous membranes (Roth/Daunderer)	Triterpene saponins	
Positive monograph, indications: digestive problems, especially as a result of dyskinesia of the biliary tract, catarrhs of the upper respiratory tract			**Commission E:** 50–100 ml pressed juice/day
Positive monograph, indications: mild, essential hypertension, **WHO monograph:** treatment of mild essential hypertension, Rauwolfia alkaloids prescription-only, R. serpentina, R. tetraphylla and R. vomitoria pharmacy-only, highly toxic ++ (Roth/Daunderer), very toxic Ib (Wink/Wyk)	Depression, fatigue, temporary impotence, contraindications and interactions with medicinal products reported (Commission E monograph), arrhythmia, bradycardia up to cardiac arrest, acidosis, shock (Roth/Daunderer)	Indole alkaloids (e.g. reserpine, ajmalin, yohimbine)	**Commission E:** 600 mg drug, equivalent to 6 mg total alkaloids/day **WHO:** 200 mg powder/day, average dose = 50–300 mg/day

Stock plant (Latin)	Plant part	F	NF	MP	trad. MP	List A	List B	List C
Rhamnus catharticus L., syn. Cervispina cathartica (L.) MOENCH, R. wicklia TEN., Rhamnus wikkor LUCÉ, Rhamnus willdenowiana HORT. ex K. KOCH	Fruit			×		×		
Rhamnus frangula L., syn. Frangula alnus MILL., Frangula frangula (L.) KARST., Frangula vulgaris BORGH.	Bark			×		×		
Rhamnus purshiana DC. syn. Frangula purshiana (DC.) COOP.	Bark			×		×		

Remarks	Risks	Critical plant substances	Pharmacologically effective dose
Positive monograph, indications: short-term treatment of constipation, **standard marketing authorisation:** 1089.99.99, buckthorn berries and preparations thereof pharmacy-only, BfArM graduated plan of 21 June 1996 concerning medicinal products for human use containing anthranoid-containing plant genera, toxic (Roth/Daunderer), toxic II (Wink/Wyk)	strong laxative, gastrointestinal complaints observed in individual cases, long-term use/abuse causes electrolyte loss (in particular potassium), may be accompanied by muscular asthenia and cardiac dysfunction, albuminuria, haematuria, pseudomelanosis coli, aggravation of colonic inertia (development of laxative addiction) resulting from paralysis of intestinal muscles due to loss of potassium, nausea, vomiting, diarrhoea, renal irritation (Roth/Daunderer), contraindications and interactions reported (Commission E monograph)	Anthranoids	**Commission E:** 20–30 mg hydroxyanthracene derivatives/day, no administration to children under the age of 12
Positive monograph for Rhamnus frangula, indications: conditions in which easy defaecation with a soft stool is desirable, e.g. anal fissures, haemorrhoids, after rectal-anal surgical procedures, constipation, **standard marketing authorisation:** 9399.99.99, **WHO monograph:** short-term treatment of occasional constipation, as a single dose, for total intestinal evacuation before X-rays and other diagnostic examinations when electrolyte solutions alone are insufficient for adequate evacuation or the use of electrolyte solutions is not possible, **ESCOP monograph:** short-term treatment of occasional constipation, **HMPC monograph:** short-term use in cases of occasional constipation, toxic II (Wink/Wyk), BfArM graduated plan of 21 June 1996 concerning medicinal products for human use containing anthranoid-containing plant genera, toxic + (Roth/Daunderer)	gastrointestinal complaints observed in individual cases, long-term use/abuse causes electrolyte loss (in particular potassium), may be accompanied by muscular asthenia and cardiac dysfunction, albuminuria, haematuria, pseudomelanosis coli, aggravation of colonic inertia (development of laxative addiction) resulting from paralysis of intestinal muscles due to loss of potassium, contraindications and interactions reported, nausea, vomiting, bloody diarrhoea, abdominal pain, nephritis, circulatory collapse, assumed to be mutagenic (Commission E monograph)	Anthranoids	**Commission E, ESCOP:** 20–30 mg hydroxyanthracene derivatives/day, calculated as glucofrangulin A **WHO:** 0.5–2.5 g drug/day, the individually correct dose is the smallest dose required to maintain a soft stool. No administration to children under the age of 10 **HMPC:** equivalent to 10–30 mg hydroxyanthracene derivatives/day (calculated as glucofrangulin) no administration to children under the age of 12
Positive monograph, indications: constipation, colonic inertia, **standard marketing authorisation:** 8699.99.99, **WHO monograph:** short-term treatment of occasional constipation, **ESCOP monograph:** short-term use in cases of occasional constipation, **HMPC monograph:** short-term use in cases of occasional constipation, toxic II (Wink/Wyk), toxic + (Roth/Daunderer), BfArM graduated plan of 21 June 1996 concerning medicinal products for human use containing anthranoid-containing plant genera	strong laxative, nausea, vomiting, bloody diarrhoea, abdominal pain, nephritis, circulatory collapse, assumed to be mutagenic (Wink/Wyk), gastrointestinal complaints observed in individual cases, long-term abuse causes electrolyte loss, in particular potassium, albuminuria and haematuria, pigment implantation into the intestinal mucosa (pseudomelanosis coli), aggravation of colonic inertia (development of laxative addiction) resulting from paralysis of intestinal muscles due to loss of potassium, contraindications and interactions reported (Commission E processing monograph)	Anthranoids, saponins	**Commission E:** 20–30 mg hydroxyanthracene derivatives/day calculated as cascaroside A **ESCOP:** 1.5–2 g drug/day, preparations equivalent to 20–30 mg hydroxyanthracene derivates (calculated as cascaroside A), no administration to children under the age of 10 **WHO:** 0.3–1 g drug/day **HMPC:** equivalent to 10–30 mg hydroxyanthracene derivatives/day (calculated as cascaroside A), no administration to children under the age of 12

Stock plant (Latin)	Plant part	F	NF	MP	trad. MP	List A	List B	List C
Rheum officinale BAILLON	Root	× F		×			×	
Rheum palmatum L., syn. Rhabarbarum palmatum (L.) MOENCH, Rheum laciniatum PRAIN	Root	× F		×			×	
Rheum rhabarbarum L., syn. Rheum franzenbachii MÜNT., Rheum muricatum BLANCO, Rheum undulatum L.	Stem	×						
Rhodiola rosea L.	Herb		Not NFS					×
Rhodiola rosea L.	Root		Not NFS		×			×
Rhododendron spp., (medicinal plant: Rhododendron ferrugineum L, syn. Chamaerhododendron ferrugineum BUBANI)	all plant parts					×		
Rhus aromatica AIT., syn. Rhus canadensis MARSH. non MILL.	Bark		×					

Remarks	Risks	Critical plant substances	Pharmacologically effective dose
Positive monograph, indications: constipation, **standard marketing authorisation:** 1189.99.99, **WHO monograph:** short-term treatment of occasional constipation, **ESCOP monograph:** short-term use in cases of occasional constipation, **HMPC monograph:** short-term use in cases of occasional constipation, BfArM graduated plan of 21 June 1996 concerning medicinal products for human use containing anthranoid-containing plant genera	contains anthranoids, disturbance of electrolyte balance (in particular potassium), albuminuria, haematuria, reflectory activation of uterus during pregnancy, transfer of aglycones into breast milk, gastrointestinal cramps observed in individual cases, contraindications and interactions reported (Schilcher)	Oxalic acid, anthranoids	**Commission E:** 20–30 mg hydroxyanthracene derivatives/ day, calculated as rhein, 30–120 mg hydroxyanthracene derivatives equivalent to 1.2 to 4.8 g drug **ESCOP:** equivalent to 15–50 mg hydroxyanthracene derivatives/ day (calculated as rhein), no administration to children under the age of 10 **WHO:** 0.5–1.5 g drug/day, equivalent to 10–30 mg hydroxyanthracene derivatives **HMPC:** equivalent to 20–30 mg hydroxyanthracene derivatives/ day (calculated as rhein), no administration to children under the age of 12
Positive monograph, indications: constipation **standard marketing authorisation:** 1189.99.99, **WHO monograph (rhizoma):** short-term treatment of occasional constipation, **ESCOP monograph:** short-term use in cases of occasional constipation, **HMPC monograph:** short-term use in cases of occasional constipation, BfArM graduated plan of 21 June 1996 concerning medicinal products for human use containing anthranoid-containing plant genera, mildly toxic III (Wink/ Wyk)	Disturbance of electrolyte balance (in particular potassium), albuminuria, haematuria, reflectory activation of uterus during pregnancy, transfer of aglycones into breast milk, gastrointestinal cramps observed in individual cases, contraindications and interactions reported (Schilcher)	Oxalic acid, anthranoids	**Commission E:** 20–30 mg hydroxyanthracene derivatives/ day, calculated as rhein, 30–120 mg hydroxyanthracene derivatives, equivalent to 1.2 to 4.8 g drug **ESCOP:** equivalent to 15–50 mg hydroxyanthracene derivatives/ day (calculated as rhein), no administration to children under the age of 10 **WHO:** 0.5–1.5 g drug/day, equivalent to 10–30 mg hydroxyanthracene derivatives **HMPC:** equivalent to 20–30 mg hydroxyanthracene derivatives/ day (calculated as rhein), no administration to children under the age of 12
slightly toxic (+) (Roth/Daunderer), mildly toxic III (Wink/Wyk)	Cases of poisoning in children observed after consumption of large amounts of raw and unripe rhubarb stalks (Roth/Daunderer)	Oxalic acid, anthranoids	
s. BfR statement: http://www.bfr.bund.de/ cm/350/risikobewertung-von-pflanzen-und-pflanzlichen-zubereitungen.pdf, **HMPC monograph:** (Rhodiolae roseae rhizoma et radix): traditional use			
Negative monograph: due to risks associated with Rhododendron ferrugineum, toxic Ib to very toxic II (Wink/Wyk), toxic + to highly toxic ++ (Roth/Daunderer)	Burning in mouth and throat, hypersalivation, vomiting, cramps, diarrhoea, also affects CNS: headache, faintness, convulsions, death by respiratory arrest (Wink/Wyk)	Triterpenes (e.g. ursolic acid), hydroquinones (e.g. arbutin), andromedotoxin (grayanotoxin I)	

Stock plant (Latin)	Plant part	F	NF	MP	trad. MP	List A	List B	List C
Rhus aromatica AIT., syn. Rhus canadensis MARSH. non MILL.	Root cortex		Not NFS					×
Rhus coriaria L., Toxicodendron coriaria KUNTZE.	Fruit	× S						
Rhus toxicodendron L., syn. Toxicodendron quercifolium (MICHX.) GREENE, Toxicodendron toxicodendron (L.) BRITT	all plant parts					×		
Ribes nigrum L., syn. Ribes olidum MOENCH	Leaf	× T		×	×		×	
Ribes nigrum L., syn. Ribes olidum MOENCH	Fruit	×						
Ribes rubrum L., syn. Ribes domesticum JANCZ., Ribes sativum (RCHB.) SYME	Fruit	×						
Ricinus communis L., syn. Croton spinosus L., Ricinus africanus MILL., Ricinus. inermis JACQ., Ricinus. vulgaris MILL.	Seed					×		
Robinia pseudoacacia L., syn. Pseudoacacia odorata MOENCH	Flower	× T, S b					×	
Rosa canina L., syn. Rosa armata STEV. ex BESSER, Rosa caucasica PALL.	Fruit	×		×			×	
Rosa gallica L., syn. Rosa grandiflora SALISB., Rosa provincialis AIT., Rosa provincialis HERRM., Rosa pumila JACQ., Rosa rubra LAM.	Petal	× T						
Rosa x centifolia, syn. Rosa centifolia x vulgaris SERINGE ex DC., Rosa gallica x centifolia REGEL	Petal	× T						

Remarks	Risks	Critical plant substances	Pharmacologically effective dose
extremely toxic Ia (Wink/Wyk)	strong irritation of mucous membranes in mouth, throat and gastrointestinal tract, nausea, vomiting, gastroenteritis accompanied by bloody diarrhoea and colic, vertigo, dizziness, excitement and serious kidney damage (Wink/Wyk)	Pyrocatechol derivatives (urushiols, toxicondendrine)	
ESCOP monograph: adjuvant in the treatment of rheumatic conditions, **HMPC monograph:** traditional use			**ESCOP:** 20–50 g leaves/litre, 250–500 ml/day, equivalent preparations
extremely toxic +++ (Roth/Daunderer), extremely toxic Ia (Wink/Wyk), **WHO monograph (oil):** short-term treatment (3–5 days) for acute constipation when other dietary methods or bulk-forming laxatives have not provided adequate relief. As a cathartic for use in bowel evacuation prior to surgery	Nausea, bloody diarrhoea, vomiting accompanied by haemorrhage, acute gastroenteritis, nephritis, liver damage, convulsions, tachycardia, circulatory arrest, death by cardiac or circulatory failure or uraemia, consumption of 6 seeds fatal to children, 10–20 seeds fatal to adults (Wink/Wyk)	Lectins (e.g. ricin), alkaloids (e.g. ricinin)	**WHO:** 1–10 ml, no administration to children under the age of 12
highly toxic ++ (Roth/Daunderer), very toxic Ib (Wink/Wyk),	all plant parts toxic, in particular seeds and fruits, contains the lectins phasin and robin, haemagglutinating properties, cytotoxic, nausea, vomiting, diarrhoea, drowsiness, mydriasis, convulsive fits, abdominal pain (Wink/Wyk), toxicity of robin is eliminated by heating (Roth/Daunderer)	Lectins (e.g. robin, phasin)	
Neutral monograph, ESCOP monograph: adjuvant treatment of pain and joint stiffness associated with osteoarthritis			**ESCOP:** 2–5 g drug, several times a day
Positive monograph: for external use			**Commission E:** 1–2 g drug/day, equivalent preparations
Positive monograph: for external use, **HMPC monograph:** in progress (external use)			

Stock plant (Latin)	Plant part	F	NF	MP	trad. MP	List A	List B	List C
Rosmarinus officinalis L., syn. Rosmarinus angustifolius MILL., Rosmarinus flexuosus JORD. et FOURR., Rosmarinus laxiflorus DENOÉ, Salvia rosmarinus SCHLEI-DEN	Leaf	× S, T		×	×		×	
Rubia tinctorum L., syn. Rubia tinctoria SALISB.	Root			×		×		
Rubus fructicosus L.	Leaf	× T		×			×	
Rubus fructicosus L.	Fruit	×						
Rubus fructicosus L.	Root		×					
Rubus idaeus L., syn. Rubus framboesianus LAMK., Rubus rosaefolius SM.	Leaf	× T						
Rubus idaeus L., syn. Rubus framboesianus LAMK., Rubus rosaefolius SM.	Fruit	×						
Rubus suavissimus S. LEE, syn. Rubus chingii var. Suavissimus S. LEE	Leaf	× T						
Rumex acetosa L., syn. Lapathum acetosa SCOP., Lapathum pratense LAM.	Herb	×						
Ruscus aculeatus L.	Rhizome		Not NFS	×	×		×	

Remarks	Risks	Critical plant substances	Pharmacologically effective dose
Positive monograph, indications: digestive problems, **standard marketing authorisation:** 1219.99.99, **WHO monograph:** carminative and spasmolytic to treat dyspepsia, traditional use, **ESCOP monograph:** improvement of hepatic and biliary function and in dyspeptic complaints, **HMPC monograph:** traditional use, slightly toxic (+) (Roth/Daunderer)	Ingestion of high doses causes coma, tonic-clonic cramps, impaired patellar reflex, reduced pupillary reflex, vomiting, gastroenteritis, uterine haemorrhage, leucocytosis, renal irritation, albuminuria, in severe cases death by pulmonary oedema (Roth/Daunderer), specific case reports on this are not available (Hager); preparations, predominantly in combination with a complex composition, formerly used for abortion; however, case reports indicating a causal connection between ingestion of rosemary and abortion are not known. Experimental studies allowing for assessing the risk are not available either (Hager)	Eucalyptol (=1,8-cineol), camphor, alpha-pinene	**Commission E:** 4–6 g drug/day **ESCOP:** 2–4 g drug/day **WHO:** 4–6 g drug/day or as infusion 2–4 g in 150 ml water, 3 times a day
Negative monograph: due to risks, BfArM graduated plan of 15 Mar 1993 concerning medicinal products containing Rubia tinctorum, substance of concern (AMK) mildly toxic II to toxic III (Wink/Wyk)	Ingestion of high doses causes digestive problems, nausea, vomiting, bloody diarrhoea, abdominal pain, nephritis, assumed to be mutagenic (Wink/Wyk)	Hydroxyanthraquinones (e.g. lucidin)	
Positive monograph, indications: non-specific, acute forms of diarrhoea, mild inflammations of the oral and pharyngeal mucosa, **standard marketing authorisation:** 1449.99.99			**Commission E:** 4.5 g drug/day, equivalent preparations
Neutral monograph			
Neutral monograph, HMPC monograph: in progress			
slightly toxic (Roth/Daunderer)	Due to the oxalic acid content, care should be taken when consuming fresh sorrel (Frohne)	Oxalic acid	
Positive monograph, indications: adjuvant treatment of chronic venous insufficiency (especially pain and feeling of heaviness in legs, night leg cramps, itching, swelling), adjuvant treatment of symptoms associated with haemorrhoids such as itching and burning, **HMPC monograph:** traditional use, **ESCOP monograph:** supportive therapy for symptoms of chronic venous insuffiency, such as painful, tired and heavy legs, tingling and swelling, supportive therapy for symptoms of haemorrhoids, such as itching and burning, slightly toxic (+) to toxic + (Roth/Daunderer), mildly toxic III (Wink/Wyk)	Gastric complaints and nausea may occur in rare cases (Commission E monograph)		**Commission E, ESCOP:** extract equivalent to 7–11 mg ruscogenin/day

Stock plant (Latin)	Plant part	F	NF	MP	trad. MP	List A	List B	List C
Ruta graveolens L., syn. Ruta graveolens L. var. vulgaris WILLK., Ruta hortensis MILL.	Leaf, herb			×		×		
Salix alba L.	Bark	× T		×	×		×	
Salvia divinorum EPLING et JÁTIVA	all plant parts					×		
Salvia officinalis L., syn. Salvia chromatica et papillosa HOFFMG., Salvia maior et minor GMELI, Salvia tomentosa MILL.	Leaf	× S, T		×	×		×	
Salvia sclarea L., syn. Aethiopis sclarea (L.) FOURR., Salvia vulgaris MILL.	Leaf	× S						
Salvia sclarea L., syn. Aethiopis sclarea (L.) FOURR., Salvia vulgaris MILL.	Flower		Not NFS					×
Salvia triloba L. FIL., syn. Salvia fruticosa MILL., Salvia lobryana AZNAV.	Leaf	× S, T						
Sambucus ebulus L., syn. Ebulum humile GARCKE	Fruit		×					
Sambucus nigra L., syn. Sambucus arborescens GILIB., Sambucus medullina GILIB.	Leaf		Not NFS					×
Sambucus nigra L., syn. Sambucus arborescens GILIB., Sambucus medullina GILIB.	Flower	×		×	×		×	

Remarks	Risks	Critical plant substances	Pharmacologically effective dose
Negative monograph: due to risks, toxic + (Roth/Daunderer), toxic Ib to very toxic II (Wink/Wyk)	Contact dermatitis, swelling of tongue, hypersalivation, severe gastroenteritis, diuresis, mydriasis, visual disturbances, haematuria, even death, uterus contractions, therefore formerly used as aborticide (Wink/Wyk)	Furocoumarins, quinoline alkaloids	
Positive monograph, indications: febrile diseases, rheumatic conditions, headache, **WHO monograph:** symptomatic treatment of fever and pain, and symptomatic treatment of mild rheumatic conditions, treatment of the common cold, traditional use, **ESCOP monograph:** relief of low back pain, symptomatic relief of mild osteoarthritc and rheumatic complaints, **HMPC monograph:** short-term treatment of low back pain, traditional use	gastrointestinal complaints, allergic or pseudoallergic reactions to the salicylates contained in the drug (Wichtl, Schilcher), Reye's syndrome	Salicylates	**Commission E:** 60–120 mg total salicin/day **ESCOP, WHO:** equivalent to 120–240 mg salicin **HMPC:** 1572 mg extract/day on indication of well-established use, 260–500 mg comminuted drug, 3 times a day on indication of traditional use
toxic II, psychotropic (Wink/Wyk), non-marketable narcotic acc. to Annex 1 of the Narcotic Drugs Act (BtMG)	Psychotropic, hallucinogenic (Wink/Wyk), sweats, headache, disorientation, psychosis, anxiety and panic attacks reported (www.erowid.org)	Diterpenes (e.g. salvinorin A)	
Positive monograph, indications: digestive problems, excessive sweat secretion, **ESCOP monograph:** inflammations and infections of the mouth and throat such as stomatitis, gingivitis and pharyngitis, hyperhidrosis, **HMPC monograph:** traditional use, **WHO monograph:** treatment of influenza, bronchitis, sinusitis, meningitis and neuritis, **standard marketing authorisation:** 1229.99.99, slightly or hardly toxic (Roth/Daunderer)	Long-term use of alcoholic extracts and pure essential oil may cause epileptiform cramps (Commission E monograph)	Thujone, eucalyptol (=1,8-cineol), camphor	**Commission E:** 4–6 g drug/day, equivalent preparations **ESCOP:** 1–1.5 g drug in 150 ml water, also several times a day **WHO:** 4 g/day
		Eucalyptol (=1,8-cineol), camphor, estragole (=methyl chavicol)	
		Eucalyptol (=1,8-cineol), camphor, estragole (=methyl chavicol)	
		Thujone, eucalyptol (=1,8-cineol), camphor	
mildly toxic III (Wink/Wyk), toxic + (Roth/Daunderer)	Nausea, vomiting, diarrhoea, tickle in the throat, vertigo, headache, visual disturbances, cardiac complaints, bloody diarrhoea, cases of fatal poisoning also known (Wink/Wyk)	Iridoid glycosides (e.g. ebuloside), lectins	
mildly toxic III (Wink/Wyk), hardly toxic (+)	Ingestion of higher doses causes vomiting and diarrhoea (Wink/Wyk)	cyanogenic glycosides (e.g. sambunigrin), lectins	
Positive monograph, indications: coughs and colds, **standard marketing authorisation:** 1019.99.99, **WHO monograph:** traditional use, **HMPC monograph:** traditional use			**Commission E:** 10–15 g drug/day, equivalent preparations **WHO:** 3–5 g drug, 3 times a day

Stock plant (Latin)	Plant part	F	NF	MP	trad. MP	List A	List B	List C
Sambucus nigra L., syn. Sambucus arborescens GILIB., Sambucus medullina GILIB.	Fruit	× b						
Sanicula europaea L., syn. Astrantia diapensia SCOP., Caucalis capitata SALISB., Sanicula officinalis GOUAN, Sanicula trilobata GILIB., Sanicula vulgaris (KOCH) LANDOZ.	Herb		Not NFS	×			×	
Santalum album L., syn. Sandalum album RUMPH., Santalum myrtifolium (L.) ROXB., Sirium myrtifolium L.	Wood		Not NFS	×			×	
Saponaria officinalis L., syn. Bootia nervosa GILIB., Lychnis officinalis (L.) SCOP., Saponaria hybrida MILL., Saponaria nervosa GILIB., Silene officinarum RUPR., Silene saponaria FRIES.	Root	×		×			×	
Sassafras spp.	all plant parts					×		
Satureja biflora (Buch.-Ham. ex D. Don) Briq.	Herb		×					
Satureja hortensis L., syn. Clinopodium hortense O. KUNTZE, Satureja brachiata STOKES, Satureja officinarum CRANTZ,	Herb	× S						
Schinus molle L., syn. Schinus molle var. areira (L.) DC., Schinus areira L. and Schinus terebinthifolius RADDI, syn. Schinus antiarthriticus MART. ex MARCH, Schinus mellisii ENGL., Schinus mucronulata MART.	Fruit	× S					×	
Schisandra chinensis (TURCZ.) BAILL., syn. Idesia polycarpa MORR. et DEVOS, Kadsura chinensis TURCZ., Maximowiczia amurensis RUPR.	Fruit	×		×			×	

Remarks	Risks	Critical plant substances	Pharmacologically effective dose
Use ripe fruits, heat sufficiently (cook) before consumption, **HMPC monograph:** in progress	Unripe fruits, seeds and green plant parts contain sambunigrin and lectin which can be deactivated by thermal exposure (Ternes), ingestion of higher doses of unripe, unheated fruits causes vomiting and diarrhoea (Wink/Wyk)	cyanogenic glycosides (e.g. sambunigrin), lectins	
Positive monograph, indications: catarrhs of the respiratory tract			**Commission E:** 4–6 g drug/day, equivalent preparations
Positive monograph, indications: adjuvant treatment of lower urinary tract infections, toxic + (Roth/Daunderer), toxic II (Wink/Wyk)	Ingestion of high doses may cause nephritis, digestive problems, heartburn, dyspepsia and haematuria, irritating to skin (Wink/Wyk)	Santalol	**Commission E:** 1.0–1.5 g essential oil/day, 10–20 g drug, equivalent preparations
use in foods only known as an ingredient of "halva" recipes Positive monograph, indications: catarrhs of the upper respiratory tract	Irritation of gastric mucosa observed in rare cases (Commission E monograph)	Triterpene saponins	
poisonous plant, toxic + (Roth/Daunderer), toxic II (Wink/Wyk), BfArM graduated plans of 19 Jan 1995 and 13 Aug 1997 concerning medicinal products containing Sassafras, sassafras oil is a "scheduled substance" of category 1 acc. to Article 2 a of Regulations (EC) No. 273/2004 and No. 111/2005 and is thus subject to the provisions of the Precursors Control Act (GÜG).	The toxicity is basically attributed to safrole contained in the drug, which is a genotoxic carcinogen (Hager), ingestion of higher doses causes nephritis, loss of consciousness, faintness and liver damage, irritating to skin (Wink/Wyk)	Safrole	
		Carvacrol	
Both species are commercially available. (s. Guidelines for spices and other seasoning ingredients), Schinus terebinthifolius: highly toxic (Roth/Daunderer), toxic II (Wink/Wyk). After consumption of small amounts of S. terebinthifolius as spice, the reported undesirable effects were, however, not detected with fruits imported from the Réunion Island	Schinus terebinthifolius: headache, swollen eyelids, digestive problems (Wink/Wyk)		
WHO monograph: treatment of chronic cough and asthma, diabetes, urinary tract disorders, as a general tonic for treating fatigue associated with illness			**WHO:** 1.5–6 g drug/day

Stock plant (Latin)	Plant part	F	NF	MP	trad. MP	List A	List B	List C
Schoenocaulon officinale (SCHLECHTEND. et CHAM.) A. GRAY, syn. Sabadilla officinarum BRANDT, Veratrum officinale SCHLECHTEND. et CHAM., Asagraea officinalis LINDL.	Seed			×		×		
Scopolia spp.	all plant parts			×		×		
Scutellaria baicalensis GEORGI, syn. Scutellaria macrantha FISCH.	Leaf		Not NFS					×
Scutellaria baicalensis GEORGI, syn. Scutellaria macrantha FISCH.	Root		Not NFS	×			×	
Senecio spp.	all plant parts			×		×		
Serenoa repens (BARTR.) SMALL, Sabal serrulata (MICH.) NUTALL ex SCHULT., Sabal serrulata ROEM. et SCHULT., Brahea serrulata (MICHX.) H. WENDL., Chamaerops serrulata MICHX., Serenoa serrulata HOOK.	Fruit		Not NFS	×			×	
Sideritis scardica GRISEB.	Herb	× T						

Remarks	Risks	Critical plant substances	Pharmacologically effective dose
poisonous plant, Veratrum alkaloids prescription-only, highly toxic ++ (Roth/Daunderer), extremely toxic Ia (Wink/Wyk)	The alkaloids contained in the drug act as nerve and muscle poison, burning in the mouth and throat, nausea, diarrhoea, fits of rage, bradycardia, dyspnoea, convulsions, even death, highly irritating to skin (Wink/Wyk), affects CNS, mental excitement, cramps, uncontrolled rage, paralysis of the myelencephalic centres, in low doses vertigo, darkened vision, marasmus, hypothermia, bradycardia, dyspnoea and gastrointestinal irritation, damage to striated muscles (Roth/Daunderer)	steroidal alkaloids (e.g. veratridine, cevadine, sabadine)	
Positive monograph (rhizome of Scopolia carniolica JACQ.), indications: cramps in the gastrointestinal, biliary and lower urinary tract in adults and schoolchildren, isolated alkaloids and rhizome prescription-only, highly toxic Ib (Wink/Wyk) extremely toxic +++ (Roth/Daunderer)	psychotropic, facial blush, dry mucosa, tachycardia, mydriasis, hallucinations, euphoria, disorientation, micturition difficulties, insomnia, death by respiratory arrest (Wink/Wyk, Roth/Daunderer)	Tropane alkaloids (e.g. atropine, L-hyoscyamine, scopolamine)	**Commission E:** 0.25–3.0 mg total alkaloids/day, calculated as hyoscyamine
	assumed to be hepatotoxic (Parker S. Herbal medicines, adverse reactions. The Regulatory Affairs Journal, 1994, 5:29)		
WHO monograph: treatment of fever, nausea and vomiting, acute dysentery, jaundice, coughs, carbuncles and sores, and threatened abortion	assumed to be hepatotoxic (Parker S. Herbal medicines, adverse reactions. The Regulatory Affairs Journal, 1994, 5:29)		**WHO:** 3–9 g drug/day, no administration to children under the age of 12
Negative monograph: due to risks associated with wood ragwort (Senecio fucsii), substance of concern (AMK), toxic + (Roth/Daunderer) toxic II (Wink, Wyk)	hepatotoxic, withdrawal of marketing authorisations in 1992 for drugs containing pyrrolizidine alkaloids due to risks, s. Federal Gazette of 17 June 1992	Pyrrolizidine alkaloids	
Positive monograph, indications: micturition difficulties resulting from benign stage I–II prostatic hyperplasia, **WHO monograph:** treatment of lower urinary tract symptoms (nocturia, polyuria, urinary retention) secondary to BPH stages I and II, as defined by Alken, in cases where diagnosis of prostate cancer is negative, **ESCOP monograph:** symptomatic treatment of micturition disorders (dysuria, pollakisuria, nocturia, urinary retention) in mild to moderate benign prostatic hyperplasia (BPH), i.e. stages I and II as defined by Alken, stages II and II as defined by Vahlensieck, or a comparable severity of symptoms as evaluated by the International Prostate Symptom Score and associated diagnostic tests, **HMPC monograph:** in progress	Side effects: gastric complaints observed in rare cases (Commission E monograph)		**Commission E, WHO:** 1–2 g drug or 320 mg extract/day, equivalent preparations

Stock plant (Latin)	Plant part	F	NF	MP	trad. MP	List A	List B	List C
Silybum marianum (L.) GAERTN., syn. Carduus marianus L., Carthamus maculatus LAM., Crisium, maculatum SCOP., Mariana mariana (L.) HILL., Silybum maculatum MOENCH	Fruit	× T		×			×	
Sinapis alba L., syn. Bonnania officinalis PRESL., Brassica alba RABENH., Crucifera lampsana E.H.L. KRAUSE, Eruca alba NOUL., Sinapis hispida TEN.	Seed	× S						
Smilax officinalis KUNTH in H.B.K.	Root	×					×	
Solanum dulcamara L., syn. Dulcamara flexuosa MOENCH, Solanum laxum ROYLE, Solanum lyratum THUNB.	Stem				×	×		
Solanum nigrum L. emend. MILL., syn. Solanum alpinum ZOLL. et MOR., Solanum americanum MILL., Solanum vulgare PARKINS.	all plant parts					×		
Solidago virgaurea L., syn. Amphiraphis leiocarpa DC., Dectis decurrens RAF. var. leiocarpa BENTH., Solidago cantonensis LOUR.	Herb	× T		×	×		×	
Sorbus aucuparia L., syn. Aucuparia silvestris MED.	Fruit	× b						
Spartium junceum L., syn. Sarothamnus junceus LINK, Spartianthus junceus (L.) LINK	all plant parts					×		
Spinacia oleracea L.	Leaf	×						
Stachys officinalis (L.) TREVISAN, syn. Betonia officinalis L.	Herb		Not NFS					×

Remarks	Risks	Critical plant substances	Pharmacologically effective dose
Positive monograph, indications: digestive problems, toxic liver damage, adjuvant treatment of chronic inflammatory liver diseases and liver cirrhosis, **standard marketing authorisation:** 1589.99.99, **WHO monograph:** supportive treatment of acute or chronic hepatitis and cirrhosis induced by alcohol, drugs or toxins, **HMPC monograph:** in progress, **ESCOP monograph:** toxic liver damage, supportive treatment in patients with chronic inflammatory liver conditions and hepatic cirrhosis			**Commission E, WHO:** 12–15 g drug/day, equivalent preparations **ESCOP:** equivalent to 154–324 mg silymarin/day
Positive monograph: for external use, harmful in high doses (+) (Roth/Daunderer), toxic II–III (in high doses) (Wink/Wyk)	Ingestion of high doses causes gastrointestinal complaints, nausea, vomiting, diarrhoea, severe poisoning causes central paralysis, bradycardia and bradypnoea, even coma and death (Roth/Daunderer, Wink/Wyk)	Glucosinolates	
Negative monograph: due to risks, slightly toxic (+) (Roth/Daunderer)	gastric irritation, temporary liver damage (Commission E monograph), diarrhoea accompanied by vomiting, central paralysis (Roth/Daunderer)	steroidal saponins	
Positive monograph: for external use, **HMPC monograph:** (traditional use, external use), BfArM only accepted external use in the post-authorisation procedure, highly toxic ++ (Roth/Daunderer), toxic Ib to very toxic II (Wink/Wyk)	Nausea, vomiting, mydriasis, paralysis of tongue and loss of speech, fever, visual snow, painful diarrhoea, alternating pulse, dyspnoea, cramps, apnoea (Roth/Daunderer)	steroidal alkaloid glycosides	**Commission E:** 1–3 g drug/day, equivalent preparations
highly toxic ++ (Roth/Daunderer), toxic Ib to very toxic II (Wink/Wyk)	all plant parts toxic, including unripe berries, nausea, severe vomiting, mydriasis, paralysis of tongue, loss of speech, fever, visual snow, painful diarrhoea, alternating pulse, dyspnoea, cramps, leg cramps, finally apnoea (Roth/Daunderer)	steroidal alkaloid glycosides	
Positive monograph, indications: irrigation therapy of inflammatory diseases of the lower urinary tract, prevention of urinary stones and kidney gravel, **ESCOP monograph:** irrigation of the urinary tract, especially in cases of inflammation and renal gravel, and as an adjuvant in the treatment of bacterial infections of the urinary tract, **HMPC monograph:** traditional use	abdominal complaints, contact allergies (Hager), contraindications: oedemas associated with cardiac or renal insufficiency (Commission E monograph)	Sesquiterpene lactones	**Commission E:** 6–12 g drug/day, equivalent preparations **ESCOP:** 2–3 g drug, 2–3 times a day, equivalent preparations, special dosage instructions for children
Neutral monograph: mildly toxic III (Wink/Wyk), slightly toxic (+), only ripe (mealy) fruits should be consumed. Heat/dry before consumption.	Fresh mountain ash berries contain parasorbic acid, which may cause local irritations, nausea, vomiting, diarrhoea and kidney damage. Parasorbic acid is largely destroyed or eliminated by drying or cooking the berries (Frohne/Pfänder).	Parasorbic acid, prunasin, amygdalin	
highly toxic ++ (Roth/Daunderer), very toxic Ib (Wink/Wyk)	Cases of poisoning accompanied by vomiting, kidney damage and apnoea reported (Roth/Daunderer), slightly psychotropic and hallucinogenic (Wink/Wyk)	Quinolizidine alkaloids (e.g. cytisine)	
Neutral monograph			

Stock plant (Latin)	Plant part	F	NF	MP	trad. MP	List A	List B	List C
Stevia rebaudiana (BERTONI) HEMSL., syn. Eupatorium rebaudianum BERTONI	Herb		×					
Strophantus spp.	all plant parts			×		×		
Strychnos ignatii BERG., syn. Ignatia amara L. FIL., Ignatia philippinensis BLUME	Seed			×		×		
Strychnos nux vomica L., syn. Strychnos lucida WALL., Strychnos colubrina WIGHT	Seed			×		×		
Swertia chirata BUCH.-HAMILT., syn. Agathodes chirata (DON.) GRISEB.	Herb, root		Not NFS					×
Symphytum officinale L.	Herb, root					×		
Syzygium aromaticum (L.) MERR. et L.M. PERRY, syn. Caryophyllus aromaticus L., Eugenia aromatica BAILL.	Flower bud	× S						
Syzygium cumini (L.) SKEELS, syn. Eugenia cumini (L.) DRUCE, Myrtus cumini L.	Leaf, fruit		Not NFS					×
Syzygium cumini (L.) SKEELS, syn. Eugenia cumini (L.) DRUCE, Myrtus cumini L.	Bark		Not NFS	×			×	
Syzygium cumini (L.) SKEELS, syn. Eugenia cumini (L.) DRUCE, Myrtus cumini L.	Seed		Not NFS					×
Tabebuia impetigenosa (MARTIUS ex DC.) STANDLEY, syn. Bignonia heptaphylla sensu auct., Gelseminum avellanedae (LORENTZ ex GRISEB.) KUNTZE	Bark	× T						
Tabernanthe iboga BAILL.	Root					×		
Tamarindus indica L., syn. Siliqua arabica BAUH., Tamarindus officinalis HOOK.	Fruit	×						
Tamus communis L.	all plant parts					×		

Remarks	Risks	Critical plant substances	Pharmacologically effective dose
Steviol glycosides licensed as additive (E 960)			
extremely toxic +++ (Roth/Daunderer), extremely toxic Ia (Wink/Wyk), seeds and strophanthins prescription-only, Strophantus species pharmacy-only	positive inotropic effect, bradycardia up to cardiac arrest, bloody diarrhoea, nausea, vomiting, oppression, headache and neck pain, loss of consciousness, convulsions (Wink/Wyk, Roth/Daunderer)	Cardenolides (strophanthin, strophanthidin, ouabain)	
extremely toxic +++ (Roth/Daunderer), extremely toxic Ia (Wink/Wyk) strychnine prescription-only, Ignatius bean pharmacy-only	Cramps, epileptic fits, states of anxiety, hypersalivation, death by respiratory arrest (Wink/Wyk)	Indole alkaloids (e.g. strychnine, brucine)	
Negative monograph: due to risks, extremely toxic Ia (Wink/Wyk), extremely toxic +++ (Roth/Daunderer)	Cramps, epileptic fits, states of anxiety, hypersalivation, death by respiratory arrest (Wink/Wyk), faintness, exhaustion, mental disorders, icterus, permanent blindness (Roth/Daunderer)	Alkaloids (e.g. strychnine, brucine, colubrin)	
Positive monograph: for external use only, toxic II–III (Wink/Wyk), slightly toxic (+) (Roth/Daunderer), **ESCOP monograph:** for external use, BfArM graduated plan of 5 June 1992 concerning medicinal products containing pyrrolizidine alkaloids	Pyrrolizidine alkaloids contained in the drug are hepatotoxic, mutagenic and carcinogenic. High doses inhibit peripheral nerves and may induce digestive problems. (Wink/Wyk)	Pyrrolizidine alkaloids	**Commission E, ESCOP:** external use only, equivalent preparations with 35 % root extract
Positive monograph for external use, **WHO monograph:** for external use, **HMPC assessment report**			
Positive monograph, indications: non-specific forms of diarrhoea			**Commission E:** 3–6 g drug/day, equivalent preparations
Negative monograph: due to risks	In view of safe therapeutic alternatives, use for the treatment of diabetes mellitus cannot be justified. (Commission E monograph)		
mildly toxic (Wink/Wyk)	Naphthoquinones are irritating to skin and mucous membranes, ingestion of high doses causes cytotoxic effects and digestive problems (Wink/Wyk)	Naphthoquinones (e.g. lapachol)	
highly toxic ++ (Roth/Daunderer), very toxic Ib (Wink/Wyk)	anaesthetic, psychotropic, hallucinogenic, motor disturbances, paralysis, death by cardiac failure (Wink/Wyk, Roth/Daunderer)	Indole alkaloids (e.g. ibogaine)	
toxic + (Roth/Daunderer), mildly toxic III (Wink/Wyk)	Burning in the mouth, vomiting, diarrhoea, gastroenteritis, overdose may be fatal (Wink/Wyk)	steroidal saponins, oxalate raphides	

Stock plant (Latin)	Plant part	F	NF	MP	trad. MP	List A	List B	List C
Tanacetum vulgare L., syn. Chrysanthemum vulgare (L.) BERNH., non Leucanthemum vulgare L.	Flower, herb			×		×		
Taraxacum officinale WEB. ex WIGG., syn. Taraxacum officinale (WITH.) WIGGERS, Leontodon officinale WITH., Leontodon taraxacum L.	Herb, root	×		×	×		×	
Taraxacum officinale WEB. ex WIGG., syn. Taraxacum officinale (WITH.) WIGGERS, Leontodon officinale WITH., Leontodon taraxacum L.	Flower	× T						
Teucrium chamaedrys L., syn. Chamaedrys officinalis MOENCH, Teucrium officinale LAM.	all plant parts	× F					×	
Theobroma cacao L., syn. Cacao guianensis AUBL., Cacao minus GAERTN.	Seed	×						
Theobroma cacao L., syn. Cacao guianensis AUBL., Cacao minus GAERTN.	Seed husk	× T						
Thevetia peruviana (PERS.) K. SCHUM., syn. Thevetia neriifolia JUSS. ex STEUD., Cerbera thevetia	all plant parts					×		

Remarks	Risks	Critical plant substances	Pharmacologically effective dose
Negative monograph: due to risks, toxic + (Roth/Daunderer), toxic II (Wink/Wyk), substance of concern (AMK), tansy pharmacy-only	neurotoxic properties, tonic-clonic cramps, strong tachypnoea, irregular cardiac activity, liver and kidney damage, vomiting, severe gastroenteritis, mydriasis, abortifacient effect, loss of consciousness, death (Roth/Daunderer, Wink/Wyk)	Monoterpenes (e.g. eucalyptol [=1,8-cineol], thujone, camphor)	
Taraxici herba: positive monograph, indications: lack of appetite, digestive problems such as feeling of fullness and bloating, **Taraxici radix cum herba: positive monograph, indications:** disturbance of biliary flow, stimulation of diuresis, lack of appetite, digestive problems such as feeling of fullness and bloating, **standard marketing authorisation:** 1139.99.99 (Taraxici radix cum herba), **WHO monograph** (Taraxaci radix cum herba): to stimulate diuresis, increase bile flow and stimulate appetite, and for treatment of dyspepsia, **ESCOP monograph** (Taraxaci radix): restoration of hepatic and biliary function, dyspepsia, loss of appetite, (Taraxaci folium): as an adjunct to treatments where enhanced urinary output is desireable, for example, rheumatism and the prevention of renal gravel, **HMPC monograph** (Taraxici radix cum herba and Taraxaci folium): traditional use	Latex: vomiting, abdominal pain, diarrhoea, superacid gastric complaints, contraindications reported (Roth/ Daunderer, Commission E monograph)	Sesquiterpene lactones (e.g. taraxinic acid glycosides)	**Taraxaci herba: Commission E:** 4–10 g drug, 3 times a day **ESCOP:** 4–10 g, 3 times a day **WHO:** 3–4 g comminuted drug, 3 times a day, 0.75–1 g extract Taraxaci radix cum herba: **Commission E:** 1 tbsp. cut drug in 1 cup of water as tea or 3–4 g cut or comminuted drug in 1 cup of water **ESCOP:** 3–5 g, 3 times a day **WHO:** 3–4 g comminuted drug, 3 times a day, 0.75–1 g extract
Flavourings and food ingredients with flavouring properties produced using wall germander may only be used to produce alcoholic beverages (Reg. (EC) No. 1334/2008), maximum amount of teucrin A in bitter spirits (5 mg/kg) specified in Reg. (EC) No. 1334/2008, BfArM graduated plan of 10 Aug 1992 concerning medicinal products containing Teucrium chamaedris (wall germander)	Hepatotoxic effect reported (Hager)		
Neutral monograph			
Neutral monograph			
highly toxic ++ (Roth/Daunderer), very toxic Ib (Wink/Wyk)	Approx. 4 seeds may be fatal to children, 8–10 seeds fatal to adults, numbness of tongue and throat, nausea, vomiting, faintness, cramps, disorientation, visual disturbances, very slow, fading heartbeat, bradycardia, AV block, ventricular fibrillation, death (Wink/Wyk)	cardioactive glycosides	

Stock plant (Latin)	Plant part	F	NF	MP	trad. MP	List A	List B	List C
Thuja spp.	all plant parts			×		×		
Thymus serpyllum L.s.l.	Herb	× S		×			×	
Thymus vulgaris L., syn. Thymus aestivus REUTER ex WILLK. et LANGE, Thymus durius DOD.	Herb	× S, T		×			×	
Thymus x citriodorus (PERS.) SCHREB., syn. Thymus comptus HORT., Thymus jankae HORT., Thymus lanuginosus MILL. var. citriodorum PERS., Thymus serpyllum L. var. vulgaris BENTH.	Herb	× S						
Tilia cordata MILL., syn. Tilia europaea L. γ ulmifolia L., Tilia ulmiflora SCOP. and Tilia platyphyllos SCOP., syn. Tilia europaea L. p.p., Tilia grandiflora EHRH. ex W.D.J. KOCH	Leaf	× T						
Tilia cordata MILL., syn. Tilia europaea L. γ ulmifolia L., Tilia ulmiflora SCOP. and Tilia platyphyllos SCOP., syn. Tilia europaea L. p.p., Tilia grandiflora EHRH. ex W.D.J. KOCH	Flower	× T		×	×		×	
Tilia tomentosa MOENCH., syn. Tilia argentea DESF.	Leaf	× T						
Tilia tomentosa MOENCH., syn. Tilia argentea DESF.	Flower	× T						
Tribulus terrestris L.	Fruit		Not NFS	×			×	
Trichilia catigua A. JUSS.	Bark		Not NFS					×
Trichocereus pachanoi BRITT. et ROSE, syn. Echinopsis pachanoi, Cereus peruvianus	all plant parts					×		
Trifolium pratense L., syn. Trifolium purpureum GILIB. non LOISEL	Flower	× T			×			

Remarks	Risks	Critical plant substances	Pharmacologically effective dose
Tree of life pharmacy-only, extremely toxic +++ (Roth/Daunderer), very toxic Ib (Wink/Wyk)	Mydriasis, fever, visual disturbances, headache, gastroenteritis accompanied by severe vomiting and diarrhoea, tachycardia, very intense cramps, degenerative kidney, cardiac and liver damage, death by respiratory and circulatory arrest (Wink/Wyk)	Monoterpenes (e.g. thujone)	
Positive monograph, indications: catarrhs of the upper respiratory tract		Carvacrol	**Commission E:** 4–6 g drug/day, equivalent preparations
Positive monograph, indications: symptomatic treatment of bronchitis and whooping cough, catarrhs of the upper respiratory tract, **standard marketing authorisation:** 1329.99.99, **WHO monograph:** traditional use, **HMPC monograph:** traditional use, **ESCOP monograph:** catarrh of the upper respiratory tract, bronchial catarrh and supportive treatment of pertussis, stomatitis and halitosis		Monoterpenes (e.g. camphor, carvacrol, eucalyptol [=1,8-cineol])	**Commission E, ESCOP, WHO:** 1–2 g drug/day, equivalent preparations
Neutral monograph			
Positive monograph, indications: coughs and colds, dry cough, **WHO monograph:** use for colds and cold-related coughs, dry cough, diaphoretic, antispasmodic, **HMPC monograph:** traditional use, **standard marketing authorisation:** 1129.99.99			**Commission E, WHO:** 2–4 g drug/day, equivalent preparations
Neutral monograph, HMPC assessment report			
WHO monograph: treatment of cough, headache and mastitis, traditional use, toxic II (Wink/Wyk), s. BfR statement: http://www.bfr.bund.de/cm/350/risikobewertung-von-pflanzen-und-pflanzlichen-zubereitungen.pdf	contains cytotoxic saponins and alkaloids. Harmala alkaloids may intercalate DNA and are particularly active after exposure to sunlight, photosensitivity accompanied by pronounced swelling of lips and ears (Wink/Wyk)	Furostanol and spirostanol saponins, beta-carboline alkaloids (e.g. harmala)	**WHO:** 3–6 g drug/day, no administration to children under the age of 12
individually or mixed with other barks designated as catuaba			
very toxic Ib (Wink/Wyk), mescaline contained in the drug is subject to the provisions of Annex I of the Narcotic Drugs Act (BtMG)	psychotropic, hallucinogenic, ingestion of higher doses causes gastrointestinal complaints, drop in blood pressure, bradycardia, respiratory depression, vasodilation, paralysis (Wink/Wyk)	Alkaloids (e.g. mescaline)	
WHO monograph: traditional use			**WHO:** 240–480 mg extract/day

Stock plant (Latin)	Plant part	F	NF	MP	trad. MP	List A	List B	List C
Trifolium pratense L., syn. Trifolium purpureum GILIB. non LOISEL	Leaf		Not NFS					×
Trigonella caerulea (L.) SER. var. caerulea, syn. Trigonella melilotus-caerulea (L.) A. et GR.	Herb	× S						
Trigonella foenum-graecum L., syn. Buceras foenum-graecum (L.) ALL.	Seed	× S		×	×		×	
Triticum aestivum L., syn. Triticum cereale SCHRANK, Triticum sativum LAM.	Fruit	×						
Triticum aestivum ssp. spelta L.	Fruit	×						
Tropaeolum majus L., syn. Tropaeolum elatum SALISB.	Herb	×			×			
Turnera diffusa WILLD. et SCHULTES (Turnera microphylla DESV.) and Turnera aphrodisiaca (LESTER f.) WARD, [Turnera diffusa WILLD. var. aphrodisiaca (WARD) URBAN]	Leaf	×			×			
Tussilago farfara L.	Leaf			×		×		
Tussilago farfara L.	Flower, root					×		
Uncaria tomentosa (WILLD.) DC., syn. Nauclea cinchoneae DC.	Bark		Not NFS	×			×	
Uncaria tomentosa (WILLD.) DC., syn. Nauclea cinchoneae DC.	Root		Not NFS					×

Remarks	Risks	Critical plant substances	Pharmacologically effective dose
	slight nausea and urticaria observed in very rare cases, concerning risks of isoflavones isolated from red clover s. BfR statement: http://www.bfr.bund.de/cm/343/isolierte_isoflavone_sind_nicht_ohne_risiko.pdf		
Positive monograph, indications: lack of appetite, **WHO monograph:** as an adjunct for the management of hypercholesterolaemia, and hyperglycaemia in cases of diabetes mellitus. Prevention and treatment of mountain sickness, loss of appetite, Treatment of pain, and weakness and oedema of the legs, traditional use, **ESCOP monograph:** adjuvant therapy in diabetes mellitus, anorexia, as an adjunct to a low fat diet in the treatment of mild to moderate hypercholesterolaemia, **HMPC monograph:** traditional use, **standard marketing authorisation:** 2319.99.99			**Commission E, WHO:** 6 g drug/day
slightly toxic (+) (Roth/Daunderer)			
Neutral monograph: mildly toxic III (Wink/Wyk)		Eucalyptol (=1,8-cineol), arbutin	
Positive monograph, indications: acute catarrhs of the respiratory tract accompanied by cough and hoarseness, acute, mild inflammations of the oral and pharyngeal mucosa, BfArM graduated plan of 5 June 1992 concerning medicinal products containing pyrrolizidine alkaloids	contains pyrrolizidine alkaloids with hepatotoxic, carcinogenic, partly also mutagenic and teratogenic properties (Frohne/Pfänder)	Pyrrolizidine alkaloids	
Negative monograph: due to risks	contains pyrrolizidine alkaloids with hepatotoxic, carcinogenic, partly also mutagenic and teratogenic properties (Frohne/Pfänder)	Pyrrolizidine alkaloids	
WHO monograph, indications: symptomatic treatment of arthritis, rheumatism and gastric ulcers, traditional use, **HMPC monograph:** in progress			**WHO:** 20–350 mg extract/day, no administration to children under the age of 12
	Side effects: constipation, increase in uric acid, cardiovascular complaints (Frohne, Schilcher)	pentacyclic alkaloids	

Stock plant (Latin)	Plant part	F	NF	MP	trad. MP	List A	List B	List C
Urginea maritima (L) BAKER, syn. Scilla maritima L., Drimia maritima (L.) STEARN., Ornithogalum maritimum (L.) LAM., Stellaris scilla MOENCH.	Bulb			×		×		
Urtica dioica L., syn. Urtica major KANITZ, Urtica urens maxima BLACKW.	Root		Not NFS	×	× (whole plant)		×	
Urtica dioica L., syn. Urtica major KANITZ, Urtica urens maxima BLACKW.	Fruit	× T, S			×			
Urtica dioica L., syn. Urtica major KANITZ, Urtica urens maxima BLACKW.	Herb, leaf	×		×	×		×	
Urtica urens L. syn.Urtica minor FUCHS, Urtica minor MOENCH, Urtica urens minima DOD.	Herb, leaf	×		×	×		×	
Urtica urens L. syn.Urtica minor FUCHS, Urtica minor MOENCH, Urtica urens minima DOD.	Root		Not NFS	×	×		×	
Vaccinium macrocarpon AITON	Fruit	×		×			×	

Remarks	Risks	Critical plant substances	Pharmacologically effective dose
Positive monograph, indications: mild forms of cardiac insufficiency, also when accompanied by renal insufficiency, Scilla glycosides prescription-only, sea onion itself pharmacy-only, highly toxic ++ (Roth/Daunderer), very toxic Ib (Wink/Wyk)	Nausea, vomiting, gastric complaints, diarrhoea, irregular pulse, arrhythmia, hypertension, coma, cardiac arrest, contraindications and interactions reported, narrow therapeutic range of cardioactive glycosides (Wink/Wyk, Roth/Daunderer)	cardioactive glycosides	**Commission E:** 0.1–0.5 g/day, standardised sea onion powder, equivalent preparations
Positive monograph, indications: micturition difficulties resulting from prostatic adenoma (stage I–II), **WHO monograph:** symptomatic treatment of lower urinary tract disorders (nocturia, polyuria, urinary retention) resulting from BPH stages I and II, as defined by Alken, in cases where diagnosis of prostate cancer is negative, **ESCOP monograph:** symptomatic treatment of micturition disorders (dysuria, pollakisuria, nocturia, urine retention) in benign prostatic hyperplasia (BPH) at stages I and II as defined by Alken or stages II and III as defined by Vahlensiek, **HMPC monograph:** traditional use	occasionally causes mild gastrointestinal complaints (Commission E monograph)		**ESCOP, WHO:** 4–6 g drug/day, equivalent preparations, no administration to children under the age of 12
Positive monograph, indications: adjuvant treatment of rheumatic conditions, irrigation therapy of inflammatory diseases of the lower urinary tract, prevention and treatment of kidney gravel, **standard marketing authorisation:** 8599.99.99, **ESCOP monograph:** adjuvant in the symptomatic treatment of arthritis, arthrosis and/or rheumatic conditions, **HMPC monograph:** traditional use	painful itching accompanied by urtication of skin areas that come in contact with stinging hairs (Wink/Wyk)		**Commission E, ESCOP:** 8–12 g drug/day, equivalent preparations
Positive monograph, indications: adjuvant treatment of rheumatic conditions, irrigation therapy of inflammatory diseases of the lower urinary tract, prevention and treatment of kidney gravel, **standard marketing authorisation:** 8599.99.99, **ESCOP monograph:** adjuvant in the symptomatic treatment of arthritis, arthrosis and/or rheumatic conditions, **HMPC monograph:** traditional use	painful itching accompanied by urtication of skin areas that come in contact with stinging hairs (Wink/Wyk)		**Commission E, ESCOP:** 8–12 g drug/day, equivalent preparations
Positive monograph, indications: micturition difficulties resulting from prostatic adenoma (stage I–II), **WHO monograph:** symptomatic treatment of lower urinary tract disorders (nocturia, polyuria, urinary retention) resulting from BPH stages I and II, as defined by Alken, in cases where diagnosis of prostate cancer is negative, **ESCOP monograph:** symptomatic treatment of micturition disorders (dysuria, pollakisuria, nocturia, urine retention) in benign prostatic hyperplasia (BPH) at stages I and II as defined by Alken or stages II and III as defined by Vahlensiek, **HMPC monograph:** traditional use	occasionally causes mild gastrointestinal complaints (Commission E monograph)		**Commission E:** 4–6 g drug/day, equivalent preparations **ESCOP:** 4–6 g drug/day, equivalent preparations **WHO:** 4–6 g drug/day, no administration to children under the age of 12
WHO monograph: traditional use, **ESCOP monograph:** prevention of urinary tract infections			**ESCOP:** 300–750 ml juice, 200–500 mg dry extract, 2 times a day, equivalent preparations, no administration to children **WHO:** 400–450 mg drug

Stock plant (Latin)	Plant part	F	NF	MP	trad. MP	List A	List B	List C
Vaccinium myrtillus L., syn. Myrtillus niger GILIB., Myrtillus sylvatica BUBANI	Leaf	× T						
Vaccinium myrtillus L., syn. Myrtillus niger GILIB., Myrtillus sylvatica BUBANI	Fruit	×		×			×	
Vaccinium vitis-idaea L., syn. Myrtillus exigua BUBANI, Vaccinium rubrum DULAC, Vitis idaea punctata MOENCH, Vitis-idaea punctifolia S.F. GRAY	Leaf	× T						
Vaccinium vitis-idaea L., syn. Myrtillus exigua BUBANI, Vaccinium rubrum DULAC, Vitis idaea punctata MOENCH, Vitis-idaea punctifolia S.F. GRAY	Fruit	×						
Valeriana edulis NUTT. ex TORR. et GRAY ssp. Procera (H.B.K.) F.G. MEYER, syn. Valeriana mexicana DC	Root					×		
Valeriana officinalis L.	Root	× F		×	×		×	
Vanilla planifolia ANDR., syn. Vanilla fragrans (SALISB.) AMES., Myrobroma fragrans SALISB., Vanilla aromatica S.W.	Fruit	× S						
Veratrum album L., syn. Helleborus albus GÜLDENST., Melanthium album THUNB.	Rhizome					×		

Remarks	Risks	Critical plant substances	Pharmacologically effective dose
Negative monograph: due to risks, slightly toxic (+), **HMPC monograph:** in progress	Long-term use in very high doses (more than 20 g/day) causes chronic poisoning (Schilcher)		
Positive monograph, indications: non-specific, acute diarrhoea, local treatment of mild inflammations of the oral and pharyngeal mucosa, **standard marketing authorisation:** 1009.99.99, **WHO monograph:** use for the symptomatic treatment of dysmenorrhoea associated with premenstrual syndrome, circulatory disorders in patients with capillary leakage or peripheral vascular insufficiency and ophthalmic disorders, treatment of acute diarrhoea and local irritation or inflammation of the mucous membranes of the mouth and throat, **ESCOP monograph:** (Extracts): symptomatic treatement of problems related to varicose veins, such as painful and heavy legs, (dried fruit): supportive treatment of acute, non-specific diarrhoea, **HMPC monograph:** in progress			**Commission E:** 30 g drug/day, equivalent preparations **WHO:** 20–60 g drug/day
mildly toxic III (Wink/Wyk)	contains 5–8 % valepotriates, valepotriates are suspected to have an alkyleting, cytotoxic and mutagenic effect (Hager)	Valepotriates	
Positive monograph, indications: agitation, nervous sleep disturbances, **standard marketing authorisation:** 6199.99.99 (root), 6099.99.99 (tincture), **WHO monograph:** mild sedative and sleep-promoting agent, used in the treatment of states of nervous excitation and anxiety-induced sleep disturbances, **ESCOP monograph:** relief of temporary mild nervous tension and/or difficulty in falling asleep, **HMPC monograph:** relief of mild nervous tension and sleep disorders, traditional use, hardly toxic (+) (Roth/Daunderer), mildly toxic III (Wink/Wyk), used to produce spirits	sedative properties, ingestion of higher doses causes headache, agitation, sleep disturbances, arrhythmia, central paralysis, cardiac arrest, contains 0.1–2 % valepotriates (Wichtl)	Valepotriates	**Commission E, WHO:** 2–3 g drug, several times a day, equivalent preparations, no administration to children under the age of 12 **ESCOP:** 1–3 g drug/day, no administration to children under the age of 12 **HMPC:** 2–3 g drug, up to 4 times a day, equivalent preparations, no administration to children under the age of 12
poisonous plant, extremely toxic Ia (Wink/Wyk), highly toxic ++ (Roth/Daunderer)	Hypersalivation, sneezing, burning bitter taste in the mouth, vomiting, bloody diarrhoea, bradycardia, drop in blood pressure, vertigo, headache, tremor, hallucinations, pain, itching, anaesthesia, myoplegia, death by cardiac or respiratory arrest (Wink/Wyk)	steroidal alkaloids (protoveratrine A and protoveratrine B)	

Stock plant (Latin)	Plant part	F	NF	MP	trad. MP	List A	List B	List C
Verbascum spp.	Flower	× T		×	×		×	
Verbascum spp.	Leaf		Not NFS					×
Veronica officinalis L.	Herb	× T, S						
Viburnum lantana L.	all plant parts					×		
Viburnum opulus L., syn. Opulus glandulosus MÖNCH, Opulus vulgaris BORKH., Viburnum lobatum LAM.	all plant parts					×		
Vinca minor L., syn. Pervinca minor (L.) SCOPOLI, Vinca ellipticifolia STOKES	Herb			×		×		
Viola odorata L.	Flower	× C						
Viola odorata L.	Herb, rhizome		Not NFS					×
Viola tricolor L., syn. Viola tricolor L. ssp. tricolor, Viola tricolor L. ssp. vulgaris (KOCH) OBORNY.	Herb	× C		×			×	
Viscum album L., syn. Viscum stellatum D. DON.	Fruit					×		
Viscum album L., syn. Viscum stellatum D. DON.	Herb					×		

Remarks	Risks	Critical plant substances	Pharmacologically effective dose
Positive monograph, indications: catarrhs of the respiratory tract, **HMPC monograph:** traditional use, **standard marketing authorisation:** 2449.99.99			**Commission E:** 3–4 g drug/day, equivalent preparations
Neutral monograph			
toxic + (Roth/Daunderer), mildly toxic III (Wink/Wyk)	Nausea, clouded consciousness, vomiting, diarrhoea, digestive tract inflammations, cardiac arrhythmia, cramps, dyspnoea, kidney damage (Wink/Wyk, Roth/Daunderer)	Viburnin, oxalates	
toxic + (Roth/Daunderer), mildly toxic III (Wink/Wyk)	Nausea, clouded consciousness, vomiting, diarrhoea, digestive tract inflammations, cardiac arrhythmia, cramps, dyspnoea, kidney damage (Wink/Wyk, Roth/Daunderer)	Viburnin, oxalates	
Negative monograph: due to risks, toxic + (Roth/Daunderer), toxic II (Wink/Wyk), substance of concern (AMK), BfArM graduated plan of 20 July 1987 concerning medicinal products containing lesser periwinkle	strong drop in blood pressure, may be accompanied by cardiac complaints, circulatory distress and dyspnoea, gastrointestinal complaints observed in individual cases, temporary redness of skin (Roth/Daunderer), changes in blood picture such as leucopenia, lymphopenia, drop in α1-, α2- and γ-globulin levels demonstrated in animal experiments (Schilcher)	Indole alkaloids (e.g. vincamine)	
Neutral monograph			
Neutral monograph			
Positive monograph: for external use only, **ESCOP monograph** (internal and external use): skin disorders such as exzema, seborrhoea, impetigo and acne, as well as cradle cap and nappy rash of infants, **HMPC monograph:** traditional use			**ESCOP:** 1.5–4 g, 3 times a day
Negative monograph: due to risks, slightly toxic to toxic + (Roth/Daunderer), toxic II (Wink/Wyk)	Retching, raging thirst, tenesmus, bloody stool, convulsions, gastric complaints, nausea, vomiting, diarrhoea (Wink/Wyk)	Viscotoxins, lectins	
Positive monograph: homeopathic or anthroposophic medicinal product for parenteral use only, **HMPC assessment report:** slightly toxic to toxic (Roth/Daunderer), toxic (Wink/Wyk)	Shivering, hyperthermia, headache, anginal complaints, orthostatic circulatory distress, allergic reactions (Commission E, Roth/Daunderer)	Viscotoxins, lectins	

Stock plant (Latin)	Plant part	F	NF	MP	trad. MP	List A	List B	List C
Vitex agnus-castus L., syn. Agnus-castus vulgaris CARR., Vitex verticillata LAM.	Fruit		Not NFS	×			×	
Vitis vinifera L.	Leaf	×		×			×	
Vitis vinifera L.	Fruit	×						
Voacanga africana STAPF. and Voacanga thuarsii Roem-Schu	all plant parts					×		
Wisteria spp., e.g. Wisteria sinensis (SIMS) SWEET	all plant parts					×		
Withania somnifera (L.) DUN., syn. Physalis somnifera L.	Root	×		×			×	

Remarks	Risks	Critical plant substances	Pharmacologically effective dose
Positive monograph, indications: menstrual cycle irregularities, premenstrual symptoms, mastalgia, **WHO monograph:** symptomatic treatment of gynaecological disorders including corpus luteum insufficiency and hyperprolactinaemia, premenstrual syndrome, menstrual irregularities, cyclic mastalgia and also to treat hormonally-induced acne, treatment of endometrial hyperplasia and secondary amenorrhoea, endocrine-dependent dermatoses (dermatitis symmetrica dysmenorrhoica (Matzenauer-Polland syndrome)) acne vulgaris, eczema, acne rosacea), hypermenorrhoea, infertility due to hyperprolactinaemia and luteal phase defect. Used to treat fibroid cysts and infertility, to stop miscarriages due to progesterone insufficiency, to help expel the placenta after birth and also as a digestive aid, sedative, anti-infective and for the treatment of hot flushes, **ESCOP monograph:** premenstrual syndrome (PMS) including symptoms such as mastodynia or mastalgia, menstrual cycle disorders such as polymenorrhoea, oligomenorrhoea or amenorrhoea, **HMPC monograph:** treatment of premenstrual syndrome, traditional use, slightly toxic (Roth/Daunderer)	Ingestion of high doses causes dizziness, vertigo, soporific effect (Roth/Daunderer), occasionally itching and urticaria, contraindicated during pregnancy and lactation (Commission E monograph)		**Commission E, WHO:** 30–40 mg drug/day as water-alcohol extract, no administration to children under the age of 12 **ESCOP:** equivalent to 30–40 mg (240 mg) drug/day **HMPC:** 180 mg drug/day, equivalent preparations, no administration to children and adolescents under the age of 18
ESCOP monograph: (aqueous dry extract): treatment of symptoms of chronic venous insufficiency (CVI) at stages I–II as defined by Widmer, (other preparations): symptomatic treatment of problems related to varicosis such as painful and heavy legs, **HMPC monograph:** traditional use			**ESCOP:** 10 g drug in 250 ml water per day
contains psychotropic alkaloids, alkaloids contained in the drug toxic Ib to very toxic II (Wink/Wyk)	Indole alkaloids are cell and nerve poison, psychotropic effect, supraventricular tachycardia, Adam-Stokes syndrome, ventricular fibrillation, drop in blood pressure, cerebral cramps, coma, cardiac arrest (Wink/Wyk)	Indole alkaloids (e.g. voacangine, voacamine)	
toxic II (Wink/Wyk), toxic + (Roth/Daunderer)	gastric complaints accompanied by vomiting, diarrhoea, circulatory distress, facial pallor and drowsiness, headache, vertigo and circulatory collapse (Wink/Wyk) (Roth/Daunderer)	Wistarin, lectins	
mildly toxic III (Wink/Wyk), **WHO monograph:** as an antistress agent to improve reaction time, tonic to increase energy, improve overall health and prevent disease in athletes and the elderly, **HMPC monograph:** final public statement, s. BfR statement: http://www.bfr.bund.de/cm/350/risikobewertung-von-pflanzen-und-pflanzlichen-zubereitungen.pdf	Nausea, vomiting, diarrhoea (WHO monograph)	steroidal lactones (withanolides), alkaloids	**WHO:** 3–6 g drug/day or 250 mg, 2 times a day (depending on the indication)

139

Stock plant (Latin)	Plant part	F	NF	MP	trad. MP	List A	List B	List C
Xysmalobium undulatum (L.) R. BR., syn. Asclepias undulata L., Xysmalobium amplifolium WEIM., Xysmalobium dispar N. E. BR., Xysmalobium lapathifolium DECNE.	Root			×		×		
Yucca filamentosa L., syn. Yucca angustifolia hort. non PURSH, Yucca smalliana FERN.	all plant parts		Not NFS					×
Zea mays L.	Style	× T		×			×	
Zea mays L.	Seed	×						
Zingiber officinale ROSC., syn. Amomum zingiber L.	Rhizome	×		×			×	

Remarks	Risks	Critical plant substances	Pharmacologically effective dose
Positive monograph, indications: non-specific, acute forms of diarrhoea, licensed medicinal product, medicinal plant, very toxic Ib (Wink/Wyk), highly toxic ++ (Roth/Daunderer)	Since besides tannins the uzara root contains cardioactive glycosides, the specified dosage must be strictly observed. For this reason, only stan-dardised ready-to-use medicinal products should be administered (Schilcher). The effect of high doses on the heart is similar to that of digi-talis (Commission E monograph)	Glycosides with cardenolide struc-ture	**Commission E:** preparations equivalent to 1 g drug or 75 mg total glycosides, daily dose equiv-alent to 4–90 mg total glycosides, calculated as uzarin
WHO monograph: supportive treatment of chronic nephritis, treatment of cholangitis, hepatitis and cholecystitis, and of acute and chronic cystitis and urethritis. Also used in the treatment of heart disease (as a diuretic), hy-pertension, rheumatism and diabetes mellitus		Carvacrol	**WHO:** 4–8 g/day
Positive monograph, indications: diges-tive problems, prevention of symptoms of travel sickness, **WHO monograph:** prophy-laxis of nausea and vomiting associated with motion sickness, postoperative nausea, perni-cious vomiting in pregnancy, and seasickness, **ESCOP monograph:** prophylaxis of the nau-sea and vomiting of motion sickness, and as a postoperative antiemetic for minor day-case surgical procedures, **HMPC monograph:** prevention of nausea and vomiting in motion sickness, traditional use	irritating to mucous membranes (Roth/Daunderer), contraindications: vomiting in pregnancy (Commission E monograph)		**Commission E:** 2–4 g drug/day, equivalent preparations **ESCOP, WHO:** 0.5–2 g drug/day, no administration to children under the age of 6 **HMPC:** 1–2 g drug, 1 hour before start of journey, no administra-tion to children and adolescents under the age of 18

Legal basis

Regulation (EC) No. 178/2002 of the European Parliament and of the Council of 28 January 2002 laying down the general principles and requirements of food law, establishing the European Food Safety Authority and laying down procedures in matters of foods safety

Regulation (EC) No. 258/97 of the European Parliament and of the Council of 27 January 1997 concerning novel foods and novel food ingredients

Regulation (EC) No. 1925/2006 of the European Parliament and of the Council of 20 December 2006 on the addition of vitamins and minerals and certain other substances to food

Regulation (EU) No. 609/2013 of the European Parliament and of the Council of 12 June 2013 on food intended for infants and young children, food for special medical purposes, and total diet replacement for weight control and repealing Council Directive 92/52/EEC, Commission Directives 96/8/EC, 1999/21/EC, 2006/125/EC and 2006/141/EC, Directive 2009/39/EC of the European Parliament and of the Council and Commission Regulations (EC) No. 41/2009 and (EC) No. 953/2009

Commission Regulation (EC) No. 953/2009 of 13 October 2009 on substances that may be added for specific nutritional purposes in foods for particular nutritional uses

Regulation (EC) No. 1334/2008 of the European Parliament and of the Council of 16 December 2008 on flavourings and certain food ingredients with flavouring properties for use in and on foods and amending Council Regulation (EEC) No. 1601/91, Regulations (EC) No. 2232/96 and (EC) No. 110/2008 and Directive 2000/13/EC

Directive 2002/46/EC of the European Parliament and of the Council of 10 June 2002 on the approximation of the laws of the Member States relating to food supplements

Ordinance on Food Supplements of 24 May 2004 (BAnz[1] I p. 1011), last amended by Article 8 of the Regulation of 13 December 2011 (BAnz I p. 2720) (Ordinance on Food Supplements [NemV])

Food and Feed Code in the version published on 3 June 2013 (BAnz I p. 1426), amended by Article 4 Section 20 of the Law of 7 August 2013 (BAnz I p. 3154) (Foodstuffs, Consumer Goods and Feed Code [Food and Feed Code – LFGB])

Directive 2001/83/EC of the European Parliament and of the Council of 6 November 2001 on the Community code relating to medicinal products for human use

Medicinal Products Act in the version published on 12 December 2005 (BAnz I p. 3394), amended by Article 4 Section 11 of the Law of 7 August 2013 (BAnz I p. 3154) (Act on the Trade in Medicinal Products [Medicinal Products Act – AMG])

Ordinance on the Prescription of Medicinal Products of 21 December 2005 (BAnz I p. 3632), last amended by Article 1 of the Regulation of 19 February 2013 (BAnz I p. 312) (Ordinance on the Prescription of Medicinal Products [AMVV])

Ordinance on Pharmacy-Only and OTC Drugs of 24 November 1988 (BAnz I p. 2150), last amended by Article 1 of the Regulation of 21 February 2011 (BAnz I p. 314) (Ordinance on Pharmacy-Only and OTC Drugs [AMVerkRV])

Ordinance on Standard Marketing Authorisations of 3 December 1982 (BAnz I p. 1601), last amended by Article 1 of the Regulation of 19 October 2006 (BAnz I p. 2287) (Ordinance on Standard Marketing Authorisations [StandZV])

Single Convention of 1961 on Narcotic Drugs of 30 March 1961 (BAnz 1973 II p. 1354) (SuchtstÜbk)

Law of 30 August 1976 approving the Convention on Psychotropic Substances of 21 February 1971 (BAnz 1976 II p. 1477) (PsychotrStÜbkG)

---

[1] BAnz = Bundesanzeiger (Federal Law Gazette)

Narcotic Drugs Act in the version published on 1 March 1994 (BAnz I p. 358), amended by Article 4 Section 7 of the Law of 7 August 2013 (BAnz I p. 3154) (Act on the Trade in Narcotic Drugs [Narcotic Drugs Act – BtMG])

Regulation (EC) No. 273/2004 of the European Parliament and of the Council of 11 February 2004 on drug precursors

Council Regulation (EC) No. 111/2005 of 22 December 2004 laying down rules for the monitoring of trade be-tween the Community and third countries in drug precursors

Precursors Control Act of 11 March 2008 (BAnz I p. 306), amended by Article 4 Section 9 of the Law of 7 August 2013 (BAnz I p. 3154) (Act on the Monitoring of the Trade in Precursors that can be misused for the Illicit Manufacture of Narcotic Drugs [Precursors Control Act – GÜG])

References

Arzneimittelkommission der deutschen Ärzteschaft[2] (2013). Bedenkliche Rezepturarzneimittel (Unsafe compounded drugs), http://www.akdae.de/Arzneimittelsicherheit/Weitere/Bedenkliche-Rezepturarzneimittel.pdf.

Arzneimittelkommission der Deutschen Apotheker[3], Warnung von Arnikatee innerlich (Warning about ingestion of arnica tea), Pharm. Ztg., 1981, 726, 2082.

Bäumler, S. (2006). *Heilpflanzenpraxis Heute. Portraits – Rezepturen – Anwendung* (1st ed.). Amsterdam: Urban & Fischer Verlag Elsevier.

Blaschek, W., Ebel, S., Hackenthal, E., Holzgrabe, U., Keller, K., Reichling, J. & Schulz, V. (2007). *Hagers Enzyklopädie der Arzneistoffe und Drogen* (6th ed.). Stuttgart: Wissenschaftliche Verlagsgesellschaft.

Bundesinstitut für Arzneimittel und Medizinprodukte[4] (2000). BfArM warnt vor Aristolochia in Zubereitungen aus chinesischen Heilkräutern (BfArM warns about Aristolochia in preparations of Chinese herbal remedies), http://www.bfarm.de/DE/BfArM/Presse/mitteil_alt/pm01-2000.html.

Bundesinstitut für Arzneimittel und Medizinprodukte (2008). Bescheid des BfArM gem. §21 Abs. 4 AMG vom 04.03.2008 zur Einstufung eines Cistus-incanus-haltigen Produktes als Arzneimittel (BfArM notification in accordance with Section 21 Clause 4 AMG of 4 Mar 2008 on the classification of a product containing Cistus incanus as a medicinal product).

Bundesinstitut für Arzneimittel und Medizinprodukte (2008). Abwehr von Risiken durch Arzneimittel, Stufe II, Hier: Arzneimittel, die Bestandteile oder Zubereitungen aus Allium sativum enthalten, Bescheid vom 11.03.2008 (Prevention of drug-associated risks, stage II, here: Medicinal products containing constituents or preparations of Allium sativum, notification of 11 Mar 2008), http://www.bfarm.de/DE/Pharmakovigilanz/risikoinfo/stufenplanverf/Liste/stp-allium-sativum.html.

Bundesinstitut für Arzneimittel und Medizinprodukte (2010). Abwehr von Risiken durch Arzneimittel, Stufe II, Hier: Arzneimittel, die unter Verwendung von Pflanzen der Familie der Aristolochiaceae mit den Gattungen Asarum, Aristolochia, Saruma und Thottea hergestellt werden, Bescheid vom 22.07.2010 (Prevention of drug-associated risks, stage II, here: Medicinal products manufactured using plants of the Aristolochiaceae family with the genera Asarum, Aristolochia, Saruma and Thottea, notification of 22 July 2010), http://www.bfarm.de/DE/Pharmakovigilanz/risikoinfo/stufenplanverf/Liste/stp-aristolochiaceae.html; jsessionid=38873A056B112E06417AC2A6120ACBEE.1_cid322.

Bundesinstitut für Arzneimittel und Medizinprodukte (2009). Abwehr von Gefahren durch Arzneimittel, Stufe II hier: Cimicifuga-haltige Arzneimittel einschließlich homöopathischer Zubereitungen mit einer Endkonzentration bis einschließlich D2/Leberschädigungen, Bescheid vom 09.06.2009 (Prevention of drug-associated risks, stage II, here: Medicinal products containing Cimicifuga, including homeopathic preparations with a final concentration of up to and including D2/liver damage, notification of 9 June 2009), http://www.bfarm.de/SharedDocs/1_Downloads/DE/Pharmakovigilanz/stufenplverf/cimicifuga_bescheid.pdf?__blob=publicationFile.

Bundesinstitut für Arzneimittel und Medizinprodukte (1987). Abwehr von Arzneimittelrisiken, Stufe II, Immergrünkrauthaltige Arzneimittel, Bescheid vom 20.07.1987 (Prevention of drug-associated risks, stage II, medicinal products containing Vinca minor, notification of 20 July 1987).

Bundesinstitut für Arzneimittel und Medizinprodukte (2005). Abwehr von Gefahren durch Arzneimittel, Stufe II hier: Johanniskraut (Hypericum)-haltige Humanarzneimittel zur innerlichen Anwendung, Bescheid vom 10.10.2005 (Prevention of drug-associated risks, stage II, here: Medicinal products for internal use in humans containing St. John's wort (Hypericum), notification of 10 Oct 2005), http://www.bfarm.de/SharedDocs/1_Downloads/DE/Pharmakovigilanz/stufenplverf/Johanniskraut-Bescheid-051010.pdf?__blob=publicationFile.

Bundesinstitut für Arzneimittel und Medizinprodukte (2007). Abwehr von Gefahren durch Arzneimittel, Stufe II, hier: Kava-Kava (Piper methysticum)-haltige und Kavain-haltige Arzneimittel einschließlich homöopathischer Zubereitungen mit einer Endkonzentration bis einschließlich D4, Bescheid vom 21.12.2007 (Prevention of drug-associated risks, stage II, here: Medicinal products containing kava-kava (Piper methysticum) and kavain, including homeopathic preparations with a final concentration of up to and including D4, notification of 21 Dec 2007), http://www.bfarm.de/DE/Pharmakovigilanz/risikoinfo/stufenplanverf/Liste/stp-kavakava.html.

Bundesinstitut für Arzneimittel und Medizinprodukte (1992). Abwehr von Arzneimittelrisiken, Stufe II. Pyrrolizidin-Alkaloide, Bekanntmachung über die Zulassung und Registrierung von Arzneimitteln, vom 5. Juni 1992, Bescheid. BAnz. Nr.: 111, S. 4805 vom 17.6.1992 (Prevention of drug-associated risks, stage II, pyrrolizidine alkaloids, notice of the authorisation and registration of medicinal products, notification of 5 June 1992, Federal Gazette No. 111, p. 4805 of 17 June 1992).

Bundesinstitut für Arzneimittel und Medizinprodukte (2008). Abwehr von Gefahren durch Arzneimittel, Stufe II, hier: Schöllkrauthaltige Arzneimittel zur innerlichen Anwendung, Bescheid vom 08.04.2008 (Prevention of drug-associated risks, stage II, here: Medicinal products for internal use containing Chelidonium, notification of 8 April 2008), http://www.bfarm.de/DE/Pharmakovigilanz/risikoinfo/stufenplanverf/Liste/stp-schoellkraut.html.

Bundesinstitut für Arzneimittel und Medizinprodukte (2008). Humanarzneimittel zur innerlichen Anwendung ausgenommen nach HAB 1 hergestellte homöopathische Arzneimittel, welche Drogen, Drogenzubereitungen oder isolierte Inhaltsstoffe der Anthranoid-(Hydroxyanthracenderivat-)haltigen Pflanzengattungen Andira, Cassia, Rhamnus, Rheum oder Aloe enthalten (Medicinal products for internal use in humans, excluding homeopathic medicinal products manufactured according to HAB 1 which contain drugs, drug preparations or isolated ingredients of the anthranoid- (hydroxyanthracene derivative) containing plant genera Andira, Cassia,

---

[2] Drug Commission of the German Medical Association
[3] Drug Commission of German Pharmacists
[4] Federal Institute for Drugs and Medical Devices

Rhamnus, Rheum or Aloe notification of 21 June 1996), Bescheid vom 21.6.1996: BAnz Nr. 123, 5.7.1996.

Bundesinstitut für Risikobewertung (2011). Allergie durch Lupineneiweiß, Aktualisierte Stellungnahme Nr. 039/2011 des BfR vom 26. August 2011 (Allergens caused by lupine protein, updated BfR statement no. 039/2011 of 26 Aug 2011), http://www.bfr.bund.de/cm/343/allergie-durch-lupineneiweiss-in-lebensmitteln.pdf.

Bundesinstitut für Risikobewertung (2005). Backmohn ist kein Schlafmittel für Säuglinge 12/2005, 29.04.2005 (Poppy seed for baking is not a soporific for infants, 12/2005, 29 Apr 2005), http://www.bfr.bund.de/de/presseinformation/2005/12/backmohn_ist_kein_schlafmittel_fuer_saeuglinge-6279.html.

Bundesinstitut für Risikobewertung (2002). BgVV und BfArM warnen: Schwere Gesundheitsschäden durch Ephedra-Kraut (BgVV and BfArM warn: Ephedra causes serious health issues), http://www.bfr.bund.de/de/presseinformation/2002/09/bgvv_und_bfarm_warnen__schwere_gesundheitsschaeden_durch_ephedra_kraut-1007.html.

Bundesinstitut für Risikobewertung (2005). Chininhaltige Getränke können gesundheitlich problematisch sein, Aktualisierte Gesundheitliche Bewertung Nr. 020/2008 des BfR vom 17. Februar 2005 (Quinine-containing beverages may cause health problems, Updated BfR Health Assessment No 020/2008, 17 Feb 2005), http://www.bfr.bund.de/cm/343/chininhaltige_getraenke_koennen_gesundheitliich_problematisch_sein.pdf.

Bundesinstitut für Risikobewertung (2009). Die Sicherheit von Ginkgoblätter-haltigen Tees kann wegen mangelnder Daten nicht beurteilt werden (The safety of herbal teas containing ginkgo leaves cannot be evaluated due to insufficient data), http://www.bfr.bund.de/cm/343/die_sicherheit_von_ginkgoblaetter_haltigen_tees_kann_wegen_mangelnder_daten_nicht_beurteilt_werden.pdf.

Bundesinstitut für Risikobewertung (2006). Erhöhte Morphingehalte in Mohnsamen: Gesundheitsrisiko nicht ausgeschlossen 05/2006, 20.02.2006 (Elevated morphine levels in poppy seeds: Risk to health not ruled out, 05/2006, 20 Feb 2006), http://www.bfr.bund.de/de/presseinformation/2006/05/erhoehte_morphingehalte_in_mohnsamen__gesundheitsrisiko_nicht_ausgeschlossen-7409.html.

Bundesinstitut für Risikobewertung (1999). Gehalte an Cumarin, Safrol, Methyleugenol und Estragol in Lebensmitteln (insbesondere teeähnlichen Erzeugnissen), Stellungnahme des BgVV vom 27. April 1999 (Coumarin, safrole, methyl eugenol and estragole levels in food (especially tea-like products), BgVV statement of 27 Apr 1999), http://www.bfr.bund.de/cm/343/gehalte_an_cumarin_safrol_methyleugenol_und_estragol_in_lebensmitteln.pdf.

Bundesinstitut für Risikobewertung (2012). Gesundheitliche Bewertung von synephrin- und koffeinhaltigen Sportlerprodukten und Schlankheitsmitteln (Health assessment of sports and weight loss products containing synephrine and caffeine), http://www.bfr.bund.de/cm/343/gesundheitliche-bewertung-von-synephrin-und-koffeinhaltigen-sportlerprodukten-und-schlankheitsmitteln.pdf.

Bundesinstitut für Risikobewertung (2007). Isolierte Isoflavone sind nicht ohne Risiko, aktualisierte Stellungnahme Nr. 039/2007 des BfR vom 3. April 2007 (Isolated isoflavones are not without risk, updated BfR statement No. 039/2007 of 3 Apr 2007), http://www.bfr.bund.de/cm/343/isolierte_isoflavone_sind_nicht_ohne_risiko.pdf.

Bundesinstitut für Risikobewertung (2002). Minimierung von Estragol- und Methyleugenol-Gehalten in Lebensmitteln, Hintergrundpapier vom 15. Januar 2002 (Minimisation of estragole and methyl eugenol levels in foods, background paper of 15 Jan 2002), http://www.bfr.bund.de/cm/343/minimierung_von_estragol_und_methyleugenol_gehalten_in_lebensmitteln.pdf.

Bundesinstitut für Risikobewertung (2012). Neue Erkenntnisse zu Cumarin in Zimt, Stellungnahme Nr. 036/2012 des BfR vom 27. September 2012 (New insights into coumarin contained in cinnamon, BfR statement no. 036/2012 of 27 Sept 2012), http://www.bfr.bund.de/cm/343/neue-erkenntnisse-zu-cumarin-in-zimt.pdf.

Bundesinstitut für Risikobewertung (2007). Risikobewertung macahaltiger Nahrungsergänzungsmittel, Stellungnahme Nr. 024/2007 des BfR vom 3. April 2007 (Risk assessment of maca-containing food supplements, BfR statement no. 024/2007 of 3 April 2007), http://www.bfr.bund.de/cm/343/risikobewertung_macahaltiger_nahrungsergaenzungsmittel.pdf.

Bundesinstitut für Risikobewertung (2003). Verwendung von unverdünntem Teebaumöl als kosmetisches Mittel, Stellungnahme des BfR vom 1. September 2003 (Use of undiluted tea-tree oil as a cosmetic, BfR statement of 1 Sept 2003), http://www.bfr.bund.de/cm/343/verwendung_von_unverduenntem_teebaumoel_als_kosmetisches_mittel.pdf.

Bundesinstitut für Risikobewertung (2007). Verzehr von bitteren Aprikosenkernen ist gesundheitlich bedenklich, Stellungnahme Nr. 014/2007des BfR vom 03. Mai 2007 (Consumption of bitter apricot kernels presents a health risk, BfR statement no. 014/2007 of 3 May 2007), http://www.bfr.bund.de/cm/343/verzehr_von_bitteren_aprikosenkernen_ist_gesundheitlich_bedenklich.pdf.

Burger, A. & Wachter, H. (1998). *Hunnius' pharmazeutisches Wörterbuch* (8th ed.). Berlin: de Gruyter.

Committee on Toxicity of Chemicals in Food, Consumer Products and the Environment (2006). Statement on cyanogenic glycosides in bitter apricot kernels, December 2006, http://www.food.gov.uk/multimedia/pdfs/cotstatementapricot200615.pdf.

Deutschmann, F., Hohmann, B., Sprecher, E., Stahl, E., Reinhard, E. & Wagner, H. (1992). *Pharmazeutische Biologie/Drogenanalyse I: Morphologie und Anatomie* 3rd ed., Amsterdam: Urban & Fischer Verlag Elsevier.

Erowid: Documenting the Complex Relationship between Humans & Psychoactives, http://www.erowid.org/.

Europäisches Arzneibuch[5] (2011). *Europäisches Arzneibuch 7th ed. 2011 incl. Addenda 1–4. Amtliche deutsche Ausgabe (Official German Edition)* (7th ed.). Stuttgart: Deutscher Apotheker Verlag.

European Food Safety Authority (2012). Compendium of botanicals reported to contain naturally occurring substances of possible concern for human health when used in food or food supplements.

European Herbal Infusions Association (EHIA) (2013). Inventory list of Herbals considered as Food.

Fintelmann, V. & Weiss, R. F. (2009). *Lehrbuch der Phytotherapie* (12th ed.). Stuttgart: Hippokrates (Georg Thieme).

Fleischhauer, S. G. (2008). *Enzyklopädie der essbaren Wildpflanzen* (5th ed.). Aarau: AT Verlag AZ Fachverlage.

Franke, W. (1992). *Nutzpflanzenkunde. Nutzbare Gewächse der gemäßigten Breiten, Subtropen und Tropen* (5th ed.). Stuttgart: Georg Thieme.

Frohne, D. (2006). *Heilpflanzenlexikon. Ein Leitfaden auf wissenschaftlicher Grundlage* (8th ed.). Stuttgart: Wissenschaftliche Verlagsgesellschaft.

Frohne, D. & Pfänder, H. J. (2004). *Giftpflanzen. Ein Handbuch für Apotheker, Ärzte, Toxikologen und Biologen* (5th ed.). Stuttgart: Wissenschaftliche Verlagsgesellschaft.

Gerhardt, U. (1994). *Gewürze in der Lebensmittelindustrie. Eigenschaften – Technologien – Verwendung* (2nd ed.). Hamburg: Behr's.

Giftzentrale Bonn[6] (2013). Informationen über Pflanzen (Information on plants), http://www.gizbonn.de/index.php?id=284.

Hänsel, R. & Sticher, O. (2010). *Pharmakognosie – Phytopharmazie* (9th ed.). Berlin: Springer.

Herrmann, K. (1983). *Exotische Lebensmittel. Inhaltsstoffe und Verwendung, Für Biologen, Chemiker, Mediziner und Hobby-Köche* (1st ed.). Berlin: Springer.

Hiller, K. & Melzig, M. (2003). *Lexikon der Arzneipflanzen und Drogen* (1st ed.). Spektrum Akademischer Verlag Heidelberg.

---

[5] European Pharmacopoeia
[6] Poison Control Centre, Bonn

Jänicke, C., Grünwald, J. & Brendler, T. (2003). *Handbuch Phytotherapie* (1st ed.). Stuttgart: Wissenschaftliche Verlagsgesellschaft.

Klenow, S., Latté KP, Wegewitz, U., Dusemund, B., Pöting, A., Appel, K. E., Großklaus, R., Schumann, R. & Lampen, A. (2012). Risikobewertung von Pflanzen und pflanzlichen Zubereitungen, BfR Wissenschaft (1/2012), http://www.bfr.bund.de/cm/350/risikobewertung-von-pflanzen-und-pflanzlichen-zubereitungen.pdf.

Leistner, E., Breckle, S. W., Drewke, G., Krippeit-Drews, P. & Keusgen, M. (2008). *Pharmazeutische Biologie kompakt. Grundlagen – Systematik – Humanbiologie* (7th ed.). Stuttgart: Wissenschaftliche Verlagsgesellschaft.

Leitsätze für Fruchtsäfte (Guidelines for fruit juices), Neufassung (revised version) vom 27.11.2002 (Beilage Nr. 46b zum BAnz vom 7. März 2003, GMBl Nr. 8–10 S. 151 vom 20. Februar 2003).

Leitsätze für Gemüseerzeugnisse (Guidelines for vegetable products), Neufassung (revised version) vom 8.01.2008 (Beilage zum BAnz Nr. 89 vom 18.06.2008, GMBl Nr. 23–25 S. 451 ff. vom 19.06.2008.

Leitsätze für Obsterzeugnisse (Guidelines for fruit products), Neufassung (revised version) vom 8.01.2008 (Beilage zum BAnz Nr. 89 vom 18.06.2008, GMBl Nr. 23–25 S. 451 ff. vom 19.06.2008).

Leitsätze für Ölsamen und daraus hergestellte Massen und Süßwaren (Guidelines for oilseeds as well as mixtures and sweets produced thereof), vom 27.01.1965 (Beilage [Supplement] zum BAnz. 101 vom 2.6.1965, GMBl Nr. 17 S. 165 vom 23.06.1965, zuletzt geändert am 08.01.2010, BAnz. Nr. 16 vom 29.01.2010, GMBl. Nr. 5/6, S. 120 ff. vom 04.02.2010).

Leitsätze für Tee, teeähnliche Erzeugnisse, deren Extrakte und Zubereitungen (Guidelines for tea, tea-like products, extracts thereof and preparations) (Neufassung [revised version] vom 2.12.1998 [BAnz Nr. 66a vom 9.04.1999, GMBl Nr. 11 S. 228 vom 26.04.1999]).

Lieberei, R. & Reisdorff, C. (2007). *Nutzpflanzenkunde* (7th ed.). Georg Thieme Verlag Stuttgart.

Lück, E. (2004). *Von Abalone bis Zuckerwurz, Exotisches für Gourmets, Hobbyköche und Weltenbummler* (2nd ed.). Berlin: Springer.

Mueller, M. S. & Mechler, E. (2005). Medicinal Plants in Tropical Countries: Traditional Use – Experience Facts, Stuttgart: Georg Thieme.

Oetker, A. (1967). *Dr. Oetker Warenkunde Lexikon* (9th ed.). Bielefeld: Ceres.

Oetker, A. (1989). *Dr. Oetker Lexikon Lebensmittel und Ernährung* (3rd ed.) Bielefeld: Ceres.

Oetker, A. (2004). *Lebensmittel Lexikon* (4th ed.). Bielefeld: Dr. Oetker Verlag.

Pilegaard, K., Eriksen, F. D., Soerensen, M. & Gry, J. (2007). EuroFIR-NETTOX Plant List, European Food Information Resource Consortium (EuroFIR).

Pschyrembel (2006). *Pschyrembel. Naturheilkunde und alternative Verfahren* (3rd ed.) Berlin: de Gruyter.

Rätsch, C. (1998). *Enzyklopädie der psychoaktiven Substanzen Botanik, Ethnopharmakologie und Anwendungen.* Stuttgart: Wissenschaftliche Verlagsgesellschaft.

Rätsch, C. (1999). *Enzyklopädie der psychoaktiven Pflanzen. Botanik, Ethnopharmakologie und Anwendungen* (4th ed.). Aarau: AT Verlag AZ Fachverlage.

Rätsch, C. & Müller-Ebeling, C. (2003). *Lexikon der Liebesmittel. Pflanzliche, mineralische, tierische und synthetische Aphrodisiaka,* Lizenzausgabe (1st ed.). Stuttgart: Deutscher Apotheker Verlag/Aarau: AT Verlag AZ Fachverlage.

Reinhard, E., Dingermann, T., Kreis, W., Rimpler, H. & Zündorf, I. (2008). *Reinhard Pharmazeutische Biologie 1* (7th ed.). Stuttgart: Wissenschaftliche Verlagsgesellschaft.

Roth, L., Daunderer, M. & Kormann, K. (2006). *Giftpflanzen-Pflanzengifte. Vorkommen – Wirkung – Therapie – Allergische und phototoxische Reaktionen* (4th ed.). Hamburg: Nikol.

Schenk, E. G. & Naundorf, J. (1966). *Manualia Nicolai. Lexikon der tropischen, subtropischen und mediterranen Nahrungs- und Genussmittel* Herford: Nicolaische Verlagsbuchhandlung.

Schilcher, H., Kammerer, S. & Wegener, T. (2010). *Leitfaden Phytotherapie* (3rd ed.). Amsterdam: Urban & Fischer Verlag Elsevier.

Schneider, G., Dingermann, T., Hiller, K. & Zündorf, I. (2004). *Schneider Arzneidrogen* (5th ed.). Heidelberg: Spektrum.

Schrott, E., Hermann, P. & Ammon, T. (2011). *Heilpflanzen der ayurvedischen und westlichen Medizin. Eine Gegenüberstellung* (1st ed.). Berlin: Springer.

Schulz, V. & Hänsel, R. (2004). *Rationale Phytotherapie. Ratgeber für Ärzte und Apotheker* (5th ed.). Berlin: Springer.

Scientific Committee on Food (2001 a). Opinion of the Scientific Committee on Food on Methyleugenol (4-Allyl-1,2-dimethoxybenzene). Scientific Committee on Food SCF/CS/FLAV/FLAVOUR/6 ADD2 FINAL (adopted on 26 September 2001) http://europa.eu.int/comm/food/fs/sc/scf/outcome_en.html#opinions.

Scientific Committee on Food (2001 b). Opinion of the Scientific Committee on Food on Estragole (1-Allyl-4-methoxybenzene). Scientific Committee on Food SCF/CS/FLAVOUR/4 ADD1 FINAL (adopted on 26 September 2001) http://europa.eu.int/comm/food/fs/sc/scf/outcome_en.html#opinions.

Seidemann, J. (1993). *Würzmittel-Lexikon* (1st ed.). Hamburg: Behr's.

Siewek, F. (1990). *Exotische Gewürze. Herkunft, Verwendung, Inhaltstoffe.* Basel: Birkhäuser.

Ternes, W., Täufel, A., Tunger, L. & Zobel, M. (1993). *Lebensmittel-Lexikon* (3rd ed.). Hamburg: Behr's.

Ternes, W., Täufel, A., Tunger, L. & Zobel, M. (2005). *Lexikon der Lebensmittel und der Lebensmittelchemie* (4th ed.). Stuttgart: Wissenschaftliche Verlagsgesellschaft.

Teuscher, E. (2003): *Gewürzdrogen. Handbuch der Gewürze, Gewürzkräuter, ihrer ätherischen Öle und der Gewürzmischungen.* Stuttgart: Wissenschaftliche Verlagsgesellschaft.

Teuscher, E., Melzig, M. F. & Lindequist, U. (2004). *Biogene Arzneimittel. Ein Lehrbuch der Pharmazeutischen Biologie* (6th ed.). Stuttgart: Wissenschaftliche Verlagsgesellschaft.

The European Botanical Forum (2009). Review and consolidation of European Negative List, Prohibited plants in selected EU and Candidate Countries, http://www.botanicalforum.eu/uploads/EBF.TFG4.EU%20neg.list.version.2.pdf.

van Wyk, B. E., Wink, C. & Wink, M. (2004). *Handbuch der Arzneipflanzen* (1st ed.). Stuttgart: Wissenschaftliche Verlagsgesellschaft.

van Wyk, B. E. (2005). *Handbuch der Nahrungspflanzen* (1st ed.). Stuttgart: Wissenschaftliche Verlagsgesellschaft.

von Bruchhausen, F., Ebel, S., Frahm, A. W., Hackenthal, S., Hänsel, R., Holzgrabe, U., Keller, K., Nürnberg, E., Rimpler, H., Schneider, G., Surmann, P., Wolf, H. U. & Wurm, G. (1992–1998). *Hagers Handbuch der pharmazeutischen Praxis* (5th ed.). Berlin: Springer.

Wagner, H. & Wiesenauer, M. (2003). *Phytotherapie. Phytotherapie und pflanzliche Homöopathika* (2nd ed.). Stuttgart: Wissenschaftliche Verlagsgesellschaft.

Wichtl, M. (2008). *Teedrogen und Phytopharmaka. Ein Handbuch für die Praxis auf wissenschaftlicher Grundlage* (5th ed.). Stuttgart: Wissenschaftliche Verlagsgesellschaft.

Wichtl, M. (1989). *Teedrogen. Ein Handbuch für die Praxis auf wissenschaftlicher Grundlage* (2nd ed.). Stuttgart: Wissenschaftliche Verlagsgesellschaft.

Wikipedia: The Free Encyclopedia, http://de.wikipedia.org.

Wink, M., van Wyk, B. E. & Wink, C. (2008). *Handbuch der giftigen und psychoaktiven Pflanzen* (1st ed.). Stuttgart: Wissenschaftliche Verlagsgesellschaft.

Wirtschaftsvereinigung Kräuter- und Früchtetees e.V. [Herbal and Fruit Tea Trade Association] (2013). WKF Inventarliste Lebensmitteldrogen [WKF Inventory List of Plants and Plant Parts Used for Food], http://wkf.de/Recht.1957.0.html?&L=0.

Wüstenfeld, M. & Haeseler, G. (1964). *Trinkbrandweine und Liköre* (4th ed.). Berlin: Paul Parey.

Monographs of the Federal Health Office/ Federal Institute for Drugs and Medical Devices (BGA/BfArM)

Absinthii herba (Wermutkraut): BAnz Nr.[7] 228, 5 Dec 1984.

Achillea millefolium (Schafgarbe): BAnz Nr. 22a, 1 Feb 1990.

Aconitum napellus (Blauer Eisenhut): BAnz Nr. 193a, 15 Okt 1987.

Adonis herba (Adoniskraut): BAnz. Nr. 85, 5 May 1988 and BAnz Nr. 22a, 01.02.1990 (corrected).

Agni casti fructus (Keuschlammfrüchte oder Mönchspfefferfrüchte): BAnz Nr. 226, 2 Dec 1992.

Agrimoniae herba (Odermennigkraut): BAnz Nr. 50, 13 Mar 1986 and BAnz Nr. 50, 13 Mar 1990 (corrected).

Alchemilla alpinae herba (Apfelfrauenmantelkraut): BAnz. Nr. 162, 29 Aug 1992.

Alchemilla herba (Frauenmantelkraut): BAnz. Nr. 173, 18 Sept 1986.

Allii cepae bulbus (Zwiebel): BAnz Nr. 50, 13 Mar 1986.

Allii sativi bulbus (Knoblauchzwiebel): BAnz Nr. 122, 6 July 1988.

Aloe: BAnz. Nr. 133, 21 July 1993.

Althaeae folium (Eibischblätter): BAnz Nr. 43, 2 Mar 1989.

Althaeae radix (Eibischwurzel): BAnz Nr. 43, 2 Mar 1989.

Ammeos visnagae fructus (Ammi-visnaga-Früchte): BAnz Nr. 71, 15 Apr 1994.

Anethi fructus (Dillfrüchte): BAnz Nr. 193a, 15 Okt 1987 and BAnz Nr. 50, 13 Mar 1990 (corrected).

Anethi herba (Dillkraut): BAnz Nr. 193a, 15 Okt 1987.

Angelicae fructus/- herba (Angelikafrüchte/-kraut): BAnz Nr. 101, 1 June 1990.

Angelicae radix (Angelikawurzel): BAnz Nr. 101, 1 June 1990.

Anisi fructus (Anis): BAnz Nr. 122, 6 July 1988.

Anisi stellati fructus (Sternanis): BAnz Nr. 122, 6 July 1988.

Antennariae dioicae flos (Katzenpfötchenblüten): BAnz Nr. 162, 29 Aug 1992.

Apium graveolens (Sellerie): BAnz Nr. 127, 12 July 1991.

Armoraciae rusticanae radix (Meerrettich): BAnz Nr. 85, 5 May 1988.

Arnicae flos (Arnikablüten): BAnz Nr. 228, 5 Dec 1984.

Artemisia vulgaris (Beifuß): BAnz Nr. 122, 6 July 1988.

Asparagi rhizoma (Spargelwurzelstock): BAnz Nr. 127, 12 July 1991.

Atropa belladonna (Tollkirsche): BAnz Nr. 223, 30 Nov 1985.

Aurantii flos (Pomeranzenblüten): BAnz Nr. 128, 14 July 1993.

Aurantii pericarpium (Pomeranzenschale): BAnz Nr. 193a, 15 Okt 1987 and BAnz Nr. 50, 13 Mar 1990 (corrected).

Avenae fructus (Haferfrüchte): BAnz Nr. 85, 5 May 1988.

Avenae herba (Haferkraut): BAnz Nr. 193a, 15 Okt 1987.

Barosmae folium (Buccoblätter): BAnz Nr. 22a, 1 Feb 1990.

Basilici herba (Basilienkraut): BAnz Nr. 54, 18 Mar 1992.

Berberis vulgaris (Berberitze): BAnz Nr. 43, 2 Mar 1989.

Betulae folium (Birkenblätter): BAnz Nr. 50, 13 Mar 1986.

Boldo folium (Boldoblätter): BAnz Nr. 76, 23 Apr 1987 and BAnz Nr. 164 vom 1 Sept 1990 (corrected).

Borago (Boretsch): BAnz Nr. 127, 12 July 1991.

Bryoniae radix (Zaunrübenwurzel): BAnz Nr. 122, 6 July 1988.

Bursae pastoris herba (Hirtentäschelkraut): BAnz Nr. 173, 18 Sept 1986 and BAnz Nr. 50, 13 Mar 1990 (corrected).

Cacao semen (Kakaosamen): BAnz Nr. 40, 27 Feb 1991.

Cacao testes (Kakaoschalen): BAnz Nr. 40, 27 Feb 1991.

Calendulae flos (Ringelblumenblüten): BAnz Nr. 50, 13 Mar 1986.

Calluna vulgaris (Heidekraut): BAnz. Nr. 101, 1 June 1990.

Capsicum (Paprika): BAnz Nr. 22a, 1 Feb 1990.

Cardamomi fructus (Kardamom): BAnz Nr. 223, 30 Nov 1985 and BAnz Nr. 50, 13 Mar 1990 and BAnz Nr.164, 1 Sept 1990 (corrected).

Cardui mariae fructus (Mariendistelfrüchte): BAnz Nr. 50, 13 Mar 1986.

Caricae fructus (Feigen): BAnz Nr. 101, 1 June 1990.

Caricae papayae folium (Melonenbaumbläetter): BAnz Nr. 193a, 15 Okt 1987.

Caricis rhizoma (Sandriedgraswurzelstock): BAnz Nr. 101, 1 June 1990.

Carvi fructus (Kümmel): BAnz Nr. 22a, 1 Feb 1990.

Caryophylli flos (Gewürznelken): BAnz Nr. 223, 30 Nov 1985.

Centaurea cyanus (Kornblume): BAnz Nr. 43, 2 Mar 1989.

Centaurii herba (Tausendgueldenkraut): BAnz Nr. 122, 6 July 1988 and BAnz Nr. 50, 13 Mar 1990 (corrected).

Chamomillae romanae flos (Römische Kamillenblüten): BAnz Nr. 221, 25 Nov 1993.

Chelidonii herba (Schöllkraut): BAnz Nr. 90, 15 May 1985.

Chrysanthemum vulgare (Rainfarn): Banz Nr. 122, 6 July 1988.

Cichorium intybus (Wegwarte): BAnz Nr. 76, 23 Apr 1987 and BAnz Nr. 164, 1 Sept 1990 (corrected).

Cimicifugae racemosae rhizoma (Cimicifugawurzelstock): BAnz Nr. 43, 2 Mar 1989.

Cinchonae cortex (Chinarinde): BAnz Nr. 22a, 1 Feb 1990.

Cinnamomi cassiae cortex (Chinesischer Zimt): BAnz Nr. 22a, 1 Feb 1990.

Cinnamomi ceylanici cortex (Zimtrinde): BAnz Nr. 22a, 1 Feb 1990.

Cinnamomi flos (Zimtblüten): BAnz Nr. 49, 11 Mar 1992.

Citri sinensis pericarpium (Orangenschalen): BAnz Nr. 22a, 1 Feb 1990.

Cnici benedicta herba (Benediktenkraut): BAnz Nr. 193, 15 Okt 1987.

Colae semen (Kolasamen): BAnz Nr.127, 12.07.1991.

Colchicum autumnale (Herbstzeitlose): BAnz Nr. 173, 18 Sept 1986.

Colocynthidis fructus (Koloquinthen): BAnz Nr. 164, 1 Sept 1990.

Condurango cortex (Condurangorinde): BAnz Nr. 193a, 15 Okt 1987 and BAnz Nr. 50, 13 Mar 1990 (corrected).

Convallariae herba (Maiglöckchenkraut): BAnz Nr. 76, 23 Apr 1987 and BAnz Nr. 22a, 1 Feb 1990.

Coriandri fructus (Koriander): BAnz Nr. 173, 18 Sept 1986.

Crataegi flos (Weissdornblüten): BAnz Nr. 133, 19 July 1994.

Crataegi folium (Weissdornblätter): BAnz Nr. 133, 19 July 1994.

Crataegi folium cum flore (Weissdornblätter mit Blüten): BAnz Nr. 133, 19 July 1994.

Crataegi fructus (Weissdornfrüchte): BAnz Nr. 133, 19 July 1994.

Croci stigma (Safran): BAnz Nr. 76, 23 Apr 1987.

Cucurbitae peponis semen (Kürbissamen): BAnz Nr. 223, 30 Nov 1985 and BAnz Nr. 11, 17 Jan 1991 (corrected).

Curcumae longae rhizoma (Curcumawurzelstock): BAnz Nr. 223, 30 Nov 1985 and BAnz Nr.164, 1 Sept 1990 (corrected).

Curcumae xanthorrhizae rhizoma (Javanische Gelbwurz): BAnz Nr. 122, 6 July 1988 and BAnz Nr. 164, 1 Sept 1990 (corrected).

Cymbopogon species (Cymbopogon-Arten): BAnz Nr. 22a, 1 Feb 1990.

Cynarae folium (Artischockenblätter): BAnz Nr. 122, 6 July 1988 and BAnz Nr. 164, 1 Sept 1990 (corrected).

Cynoglossi herba (Hundszungenkraut): BAnz Nr. 43, 2 Mar 1989.

Cytisi scoparii flos (Besenginsterblüten): BAnz Nr. 11, 17 Jan 1991.

Cytisi scoparii herba (Besenginsterkraut): BAnz Nr. 11, 17 Jan 1991.

Delphinii flos (Rittersspornblüten): BAnz Nr. 80, 27 Apr 1990.

Droserae herba (Sonnentaukraut): BAnz Nr. 228, 5 Dec 1984.

Dryopteris filix-mas (Wurmfarn): BAnz Nr. 180, 24 Sept 1993.

Dulcamarae stipites (Bittersüssstengel): BAnz Nr. 101, 1 June 1990.

Echinacea angust./-pallida ((schmalblättr.) Sonnenhut): BAnz Nr. 162, 29 Aug 1992.

Echinacea-pallidae-radix (Echinacea-pallida-Wurzel): BAnz Nr. 162, 29 Aug 1992.

Echinaceae purpureae herba (Purpursonnenhutkraut): BAnz Nr. 43, 2 Mar 1989.

Echinaceae-purpurae-radix (Purpursonnenhutwurzel): BAnz Nr. 162, 29 Aug 1992.

Eleutherococci radix (Eleutherococcus-senticosus-Wurzel): BAnz Nr. 11, 17 Jan 1991.

Ephedrae herba (Ephedrakraut): BAnz Nr. 11, 17 Jan 1991.

Equiseti Herba (Schachtelhalmkraut): BAnz Nr. 173, 18 Sept 1986.

Eucalypti folium (Eucalyptusblätter): BAnz Nr. 177a, 24 Sept 1986 and BAnz Nr. 50, 13 Mar 1990 (corrected).

---

[7] BAnz = Bundesanzeiger (Federal Law Gazette); Nr. = Number (No.)

References

Euphrasia officinalis (Augentrost): BAnz Nr. 162, 29 Aug 1992.

Farfarae flos/-herba/-radix (Huflattichblüten/-kraut/-wurzel): BAnz Nr. 138, 27 July 1990.

Farfarae folium (Huflattichblätter): BAnz Nr. 138, 27 July 1990.

Filipendula ulmaria (Maedesüß): BAnz Nr. 43, 2 Mar 1989.

Foeniculi fructus (Fenchel): BAnz Nr. 74, 19 Apr 1991.

Foenugraeci semen (Bockshornsamen): BAnz Nr. 22a, 1 Feb 1990.

Fragariae folium (Erdbeerblätter): BAnz Nr. 22a, 1 Feb 1990.

Frangulae cortex (Faulbaumrinde): BAnz Nr. 133, 21 July 1993.

Fumariae herba (Erdrauchkraut): BAnz Nr. 173, 18 Sept 1986.

Galangae rhizoma (Galgantwurzelstock): BAnz Nr. 173, 18 Sept 1986 and BAnz Nr. 50, 13 Mar 1990 (corrected).

Galegae officinalis herba (Geißrautenkraut): BAnz Nr. 180, 24 Sept 1993.

Galeopsidis herba (Hohlzahnkraut): BAnz Nr. 76, 23 Apr 1987.

Galii odorati herba (Waldmeisterkraut): BAnz Nr. 193a, 15 Okt 1987.

Gelsemii rhizoma (Gelsemiumwurzelstock): BAnz Nr. 178, 21 Sept 1991.

Gentianae radix (Enzianwurzel): BAnz Nr. 223, 30 Nov 1985 and BAnz Nr. 50, 13 Mar 1990 (corrected).

Ginkgo folium (Ginkgo-biloba-Blätter): BAnz Nr. 133, 19 July 1994.

Ginseng radix (Ginsengwurzel): BAnz Nr. 11, 17 Jan 1991.

Graminis flos (Heublumen): BAnz Nr. 85, 5 May 1988.

Graminis rhizoma (Queckenwurzelstock): BAnz Nr. 22a, 1 Feb 1990.

Grindeliae herba (Grindeliakraut): BAnz Nr. 11, 17 Jan 1991.

Guajaci lignum (Guajakholz): BAnz Nr. 76, 23 Apr 1987.

Gypsophilae radix (Weisse Seifenwurzel): BAnz Nr. 101, 1 June 1990.

Hamamelidis folium et cortex (Hamamelisblätter und -rinde): BAnz Nr. 154, 21 Aug 1985 and BAnz Nr. 50, 13 Mar 1990 (corrected).

Harpagophyti radix (Suedafrikanische Teufelskrallenwurzel): BAnz Nr. 43, 2 Mar 1989 and BAnz Nr.164, 1 Sept 1990 (corrected).

Harunganae madagascariensis cortex et folium (Harongarinde und -blätter): BAnz Nr. 50, 13 Mar 1986.

Hederae helicis folium (Efeublätter): BAnz Nr. 122, 6 July 1988.

Helenii radix (Alantwurzel): BAnz Nr. 85, 5 May 1988.

Helichrysi flos (Ruhrkrautblüten): BAnz Nr. 122, 6 July 1988 and BAnz Nr. 164, 1 Sept 1990 (corrected).

Hepatici nobilis herba (Leberbluemchenkraut): BAnz Nr. 128, 14 July 1993.

Herniariae herba (Bruchkraut): BAnz Nr. 173, 18 Sept 1986.

Hibisci flos (Hibiscusblüten): BAnz Nr. 22a, 1 Feb 1990.

Hippocastani cortex/-flos (Rosskastanienrinde/-Blüten): BAnz Nr. 221, 25 Nov 1993.

Hippocastani folium (Rosskastanienblätter): BAnz Nr. 128, 14.07.1993.

Hippocastani semen (Rosskastaniensamen/Trockenextrakt aus R.-samen [DAB10]): BAnz Nr. 71, 15.04.1994.

Hyoscyami folium (Hyoscyamusblätter): BAnz Nr. 85, 5 May 1988.

Hyperici herba (Johanniskraut): BAnz Nr. 228, 5 Dec 1984 and BAnz Nr. 43, 2 Mar 1989 (corrected).

Hyssopus officinalis (Ysop): BAnz Nr. 162, 29 Aug 1992.

Iridis rhizoma (Schwertlilienwurzelstock): BAnz Nr. 221, 25 Nov 1993.

Juglandis folium (Walnussblätter): BAnz Nr. 101, 1 June 1990.

Juglandis fructus cortex (Walnussfruchtschalen): BAnz Nr. 101, 1 June 1990.

Juniperi fructus (Wacholderbeeren): BAnz Nr. 228, 5 Dec 1984.

Lamii albi flos (Weisse Taubnesselblüten): BAnz Nr. 76, 23 Apr 1987.

Lamii albi herba (Weisses Taubnesselkraut): BAnz Nr. 128, 14 July 1993.

Lavandulae flos (Lavendelblüten): BAnz Nr. 228, 5 Dec 1984 and BAnz Nr. 50, 13 Mar 1990 (corrected).

Ledi palustris herba (Sumpfporstkraut): BAnz Nr. 177a, 24 Sept 1986.

Leonuri cardiacae herba (Herzgespannkraut): BAnz Nr. 50, 13 Mar 1986.

Levistici radix (Liebstoeckelwurzel): BAnz Nr. 101, 1 June 1990.

Lini semen (Leinsamen): BAnz Nr. 228, 5 Dec 1984.

Liquiritiae radix (Suessholzwurzel): BAnz Nr. 90, 15 May 1985 and BAnz Nr. 74, 19 Apr 1991 and BAnz Nr. 50, 13 Mar 1990 and BAnz Nr.178, 21 Sept 1991 (corrected).

Lupuli strobulus (Hopfenzapfen): BAnz Nr. 228, 5 Dec 1984 and BAnz Nr. 50, 13 Mar 1990 (corrected).

Lycopi herba (Wolfstrappkraut): BAnz Nr. 22a, 1 Feb 1990.

Malvae flos (Malvenblüten): BAnz Nr. 43, 2 Mar 1989.

Malvae folium (Malvenblätter): BAnz Nr. 43, 2 Mar 1989.

Manna: BAnz Nr. 22a, 1 Feb 1990.

Marrubii herba (Andornkraut): BAnz Nr. 22a 1 Feb 1990.

Mate folium (Mateblätter): BAnz Nr. 85, 5 May 1988.

Matricariae flos (Kamillenblüten): BAnz Nr. 228, 5 Dec 1984 and BAnz Nr. 50, 13 Mar 1990 (corrected).

Meliloti herba (Steinkleekraut): BAnz Nr. 50, 13 Mar 1986 and BAnz Nr. 50, 13 Mar 1990 (corrected).

Melissae folium (Melissenblätter): BAnz Nr. 228, 5 Dec 1984 and BAnz Nr. 50, 13 Mar 1990 (corrected).

Menthae piperitae folium (Pfefferminzblätter): BAnz Nr. 223, 30 Nov 1985 and BAnz Nr. 50, 13 Mar 1990 and BAnz Nr.164, 1 Sept 1990 (corrected).

Menyanthis folium (Bitterkleeblätter): BAnz Nr. 22a, 1 Feb 1990.

Myristica fragans (Muskatnussbaum): BAnz Nr. 173, 18 Sept 1986.

Myrtilli folium (Heidelbeerblätter): BAnz Nr. 76, 23 Apr 1987.

Myrtilli fructus (Heidelbeeren): BAnz Nr. 76, 23 Apr 1987 and BAnz Nr. 50, 13 Mar 1990 (corrected).

Nasturtii herba (Brunnenkressekraut): Banz Nr. 22a, 1 Feb 1990.

Oleae folium (Olivenblätter): BAnz Nr. 11, 17 Jan 1991.

Oleandri folium (Oleanderblätter): BAnz Nr. 122, 6 July 1988 and BAnz Nr. 22a, 1 Feb 1990 and BAnz Nr. 43, 2 Mar 1989 (corrected).

Ononidis radix (Hauhechelwurzel): BAnz Nr. 76, 23 Apr 1987 and BAnz Nr. 50, 13 Mar 1990 (corrected).

Origanum majorana (Majoran): BAnz Nr. 226, 2 Dec 1992.

Orthosiphonis folium (Orthosiphonblätter): BAnz Nr. 50, 13 Mar 1986 and BAnz Nr. 50, 13 Mar 1990 (corrected).

Paeonia (Pfingstrose): BAnz Nr. 85, 5 May 1988.

Passiflorae herba (Passionsblumenkraut): BAnz Nr. 223, 30 Nov 1985 and BAnz Nr. 50, 13 Mar 1990 (corrected).

Petasites hybridus/-folium (Pestwurz/-blätter): BAnz Nr. 138, 27 July 1990.

Petasitidis rhizoma (Pestwurzwurzelstock): BAnz Nr. 138, 27 July 1990.

Petroselini fructus (Petersilienfrüchte): BAnz. Nr. 43, 2 Mar 1989.

Petroselini herba/-radix (Petersilienkraut/-wurzel): BAnz. Nr. 43, 2 Mar 1989.

Phaseoli fructus sine semine (Samenfreie Gartenbohnenhülsen): BAnz Nr. 50, 13 Mar 1986 and BAnz Nr. 50, 13 Mar 1990 (corrected).

Piceae turiones recentes (Frische Fichtenspitzen): BAnz Nr. 193a, 15 Okt 1987.

Pimpinellae herba (Bibernellkraut): BAnz Nr. 101, 1 June 1990.

Pimpinellae radix (Bibernellwurzel): BAnz Nr. 101, 1 June 1990.

Pini turiones (Kiefernsprossen): BAnz Nr. 173, 18 Sept 1986 and BAnz Nr. 50, 13 Mar 1990.

Piperis methystici rhizoma (Kava-Kava-Wurzelstock): BAnz Nr. 101, 1 June 1990.

Plantaginis lanceolatae herba (Spitzwegerichkraut): BAnz. Nr. 223, 30 Nov 1985.

Plantaginis ovatae semen (Indische Flohsamen): BAnz Nr. 22a, 1 Feb 1990 and BAnz Nr. 74, 19 Apr 1991 (corrected).

Plantaginis ovatae testa (Indische Flohsamenschalen): BAnz Nr. 22a, 1 Feb 1990 and BAnz Nr. 74, 19 Apr 1991 (corrected).

Podophylli peltati rhizoma/-resina (Podophyllumwurzelstock/-Harz): BAnz Nr. 50, 13 Mar 1986.

Polygalae radix (Senegawurzel): BAnz Nr. 50, 13 Mar 1986 and BAnz Nr. 50, 13 Mar 1990 (corrected).

Polygoni avicularis herba (Vogelknoeterichkraut): BAnz Nr. 76, 23 Apr 1987 and BAnz Nr. 50, 13 Mar 1990 (corrected).

Potentillae anserinae herba (Gänsefingerkraut): BAnz. Nr. 223, 30 Nov 1985 and BAnz Nr. 50, 13 Mar 1990 (corrected).

Primulae flos (Schlüsselblumenblüten): BAnz Nr. 122, 6 July 1988 and BAnz Nr. 50, 13 Mar 1990 (corrected).

Primulae radix (Primelwurzel): BAnz Nr. 122, 6 July 1988 and BAnz Nr. 50, 13 Mar 1990 (corrected).

149

Pruni spinosae flos (Schlehdornblüten): BAnz Nr. 101, 1 June 1990.

Pruni spinosae fructus (Schlehdornfrüchte): BAnz Nr. 101, 1 June 1990.

Psyllii semen (Flohsamen): BAnz Nr. 223, 30 Nov 1985 and BAnz Nr. 50, 13 Mar 1990 (corrected).

Ptychopetali lignum (Potenzholz): BAnz Nr. 193a, 15 Okt 1987.

Pulmonariae herba (Lungenkraut): BAnz Nr. 193a, 15 Okt 1987.

Pulsatillae herba (Küchenschellenkraut): BAnz Nr. 223, 30 Nov 1985.

Quercus cortex (Eichenrinde): BAnz Nr. 22a, 1 Feb 1990.

Raphani sativi radix (Rettich): BAnz Nr. 177a, 24 Sept 1986.

Ratanhiae radix (Ratanhiawurzel): BAnz Nr. 43, 2 Mar 1989.

Rauwolfiae radix (Rauwolfiawurzel): BAnz Nr. 173, 18 Sept 1986.

Rhamni cathartici fructus (Kreuzdornbeeren):BAnz Nr. 221, 25 Nov 1993.

Rhamni purshianae cortex (Amer.Faulbaumrinde): BAnz Nr. 133, 21 July 1993.

Rhei radix (Rhabarberwurzel): BAnz Nr. 133, 21 July 1993.

Rhododendri ferruginei folium (Rostrote Alpenrosenblätter): BAnz Nr. 164, 1 Sept 1990.

Rhoeados flos (Klatschmohnblüten): BAnz Nr. 85, 5 May 1988.

Rosae flos (Rosenblüten): BAnz Nr. 164, 1 Sept 1990.

Rosae pseudofructus cum fructibus (Hagebutten): BAnz Nr. 164, 1 Sept 1990.

Rosmarini folium (Rosmarinblätter): BAnz Nr.223, 30 Nov 1985 and BAnz Nr.221, 28 Nov 1986 and BAnz Nr. 50, 13 Mar 1990.

Rubiae tinctorum radix (Krappwurzel): BAnz Nr. 162, 29 Aug 1992.

Rubi fruticosi folium (Brombeerblätter): BAnz Nr. 22a, 1 Feb 1990.

Rubi fruticosi radix (Brombeerwurzel): BAnz Nr. 22a, 1 Feb 1990.

Rubi idaei folium (Himbeerblätter): BAnz Nr. 193a, 15 Okt 1987.

Rusci aculeati rhizoma (Mäusedornwurzelstock): BAnz Nr. 127, 12 July 1991.

Ruta graveolens (Raute): BAnz Nr. 43, 2 Mar 1989.

Sabal fructus (Sägepalmenfrüchte): BAnz Nr. 43, 2 Mar 1989 and BAnz Nr. 22a, 1 Feb 1990 and BAnz Nr. 11, 17.01.1991 (corrected).

Salicis cortex (Weidenrinde): BAnz. Nr. 228, 5 Dec 1984.

Salviae folium (Salbeiblätter): BAnz Nr. 90, 15 May 1985 and BAnz Nr. 50, 13 Mar 1990 (corrected).

Sambuci flos (Holunderblüten): BAnz Nr. 50, 13 Mar 1986.

Saniculae herba (Sanikelkraut): BAnz Nr. 177a, 24 Sept 1986 and BAnz Nr. 50, 13 Mar 1990 (corrected).

Santali albi lignum (Weisses Sandelholz): BAnz Nr. 43, 2 Mar 1989.

Santali lignum rubrum (Rotes Sandelholz): BAnz Nr. 193a, 15 Okt 1987.

Saponariae rubrae radix (Rote Seifenwurzel): BAnz Nr. 80, 27 Apr 1989.

Sarsaparillae radix (Sarsaparillewurzel): BAnz Nr. 164, 1 Sept 1990.

Scillae bulbus (Meerzwiebel): BAnz Nr. 154, 21 Aug 1985 and BAnz Nr. 43, 2 Mar 1989 (corrected).

Scopoliae rhizoma (Glockenbilsenkrautwurzelstock): BAnz Nr. 177a, 24 Sept 1986.

Senecionis herba (Fuchskreuzkraut): BAnz Nr. 138, 27 July 1990.

Sennae folium (Sennesblätter): BAnz Nr. 133, 21 July 1993.

Sennae fructus (Sennesfrüchte): BAnz. Nr. 133, 21 July 1993.

Serpylli herba (Quendelkraut): BAnz Nr. 193a, 15 Okt 1987 and BAnz Nr. 50, 13 Mar 1990 (corrected).

Sinapis albae semen (Weisse Senfsamen): BAnz Nr. 22a, 1 Feb 1990.

Solidago (Goldrute): BAnz Nr. 193a, 15 Okt 1987 and BAnz Nr. 50, 13 Mar 1990 (corrected).

Sorbi aucupariae fructus (Ebereschenbeeren): BAnz Nr. 122, 6 July 1988.

Spinaciae folium (Spinatblätter): BAnz Nr. 85, 5 May 1988.

Stramonii folium/-semen (Stramoniumblätter/-samen): BAnz Nr. 22a, 1 Feb 1990.

Strychni semen (Brechnussamen): BAnz Nr. 173, 18 Sept 1986.

Symphyti radix (Beinwellwurzel): BAnz Nr. 138, 27 July 1990.

Syzygii cumini cortex (Syzygiumrinde): BAnz Nr. 76, 23 Apr 1987.

Syzygii cumini semen (Syzygiumsamen): BAnz Nr. 76, 23 Apr 1987.

Taraxaci herba (Löwenzahnkraut): BAnz Nr. 162, 29 Aug 1992.

Taraxaci radix cum herba (Löwenzahnwurzel mit -kraut): BAnz Nr. 228, 5 Dec 1984 and BAnz Nr.164, 1 Sept 1990 (corrected).

Thymi herba (Thymiankraut): BAnz Nr. 228, 5 Dec 1984 and BAnz Nr. 226, 2 Dec 1992 and BAnz Nr. 50, 13 Mar 1990 (corrected).

Tiliae flos (Lindenbüten): BAnz Nr. 164, 1 Sept 1990.

Tiliae folium (Lindenbätter): BAnz Nr. 164, 1 Sept 1990.

Tiliae tomentosae flos (Silberlindenbüten): BAnz Nr. 164, 1 Sept 1990.

Tormentillae rhizoma (Tormentillwurzelstock): BAnz Nr. 85, 5 May 1988 and BAnz Nr. 50, 13 Mar 1990 (corrected).

Turnera diffusa (Damiana): BAnz Nr. 43, 2 Mar 1989.

Urticae herba/-folium (Brennesselkraut/-blätter): BAnz Nr. 76, 23 Apr 1987.

Urticae radix (Brennesselwurzel): BAnz Nr. 173, 18 Sept 1986 and BAnz Nr. 11, 17 Jan 1991 and BAnz Nr.43, 2 Mar 1989 and BAnz Nr.50, 13 Mar 1990 (Correcteden).

Uvae ursi folium (Bärentraubenblätter): BAnz Nr. 109, 15 June 1994.

Uzarae radix (Uzarawurzel): BAnz Nr. 164, 1 Sept 1990.

Valerianae radix (Baldrianwurzel): BAnz Nr. 90, 15 May 1985 and BAnz Nr. 50, 13 Mar 1990 (corrected).

Verbasci flos (Wollblumen): BAnz Nr. 22a, 1 Feb 1990.

Veronicae herba (Ehrenpreiskraut): BAnz Nr. 43, 2 Mar 1989.

Vincae minoris herba (Immergrünkraut): BAnz Nr. 173, 18 Sept 1986.

Viola odorata (Märzveilchen): BAnz Nr. 111, 17 June 1994.

Violae odoratae flos (Maerzveilchenblüten): BAnz Nr. 111, 17 June 1994.

Violae tricoloris herba (Stiefmütterchenkraut): BAnz Nr. 50, 13 Mar 1986.

Visci albi fructus (Mistelbeeren): BAnz Nr. 128, 14 July 1993.

Visci albi herba (Mistelkraut): BAnz Nr. 228, 5 Dec 1984.

Yohimbehe cortex (Yohimbeherinde): BAnz Nr. 193a, 15 Okt 1987 and BAnz Nr. 22a, 1 Feb 1990 (corrected).

Zedoariae rhizoma (Zitwerwurzelstock): BAnz Nr. 122, 6 July 1988.

Zingiberis rhizoma (Ingwerwurzelstock): BAnz Nr. 85, 5 May 1988 and BAnz Nr.164, 1 Sept 1990 and BAnz Nr. 50, 13 Mar 1990 (corrected).

ESCOP Monographs

ESCOP (2003). ESCOP Monographs, The Scientific Foundation for Herbal Medicinal Products (2nd ed.). Stuttgart: Georg Thieme.

ESCOP (2009). ESCOP Monographs (2nd ed.). Supplement 2009, Georg Thieme Verlag KG Stuttgart.

WHO Monographs

WHO; WHO monographs on selected medicinal plants: Volume 1 (1999).

WHO; WHO monographs on selected medicinal plants: Volume 2 (2002).

WHO; WHO monographs on selected medicinal plants: Volume 3 (2007).

WHO; WHO monographs on selected medicinal plants: Volume 4 (2009).

WHO; WHO monographs on medicinal plants commonly used in the Newly Independent States (NIS) (2010).

HMPC Monographs

European Medicines Agency (EMA), http://www.ema.europa.eu/ema/index.jsp?curl=pages/medicines/landing/herbal_search.jsp&mid=WC0b01ac058001fa1d.

Community list pursuant to Article 16f of Directive 2001/83/EC of the European Parliament and of the Council of 6 November 2001 on the Community code relating to medicinal products for human use

European Medicines Agency (EMA), http://ec.europa.eu/health/human-use/herbal-medicines/index_en.htm.

Lists of other member states

Belgium: Arreté Royal du 29 Aout 1997 relatif à la fabrication et au commerce de denrées alimentaires composées ou contenant des plantes ou préparations de plantes.
Belgium: Federale Overheidsdienst Volksgezondheid, Veiligheid van de Voedselketen en Leefmilieu: Indicative list of plants/plant parts/substances and their novel food status (non-exhaustive list), http://www.health.fgov.be/eportal/foodsafety/foodstuffs/novelfoods/index.htm#.UmAJPKxqfU4.
Denmark: Drogelisten Vurdering af planter, svampe og dele heraf anvendt i kosttilskud og urtete.
Ireland: Herbal Substances which may be acceptable for inclusion in food supplements, IMB Draft Guideline – Version 5.1 20/04/2011.
Ireland: Herbal Substances not permitted in foods or food supplements, IMB Draft Guideline – Version 5.1, 20/4/2011.

Italy: Ministerio Della Salute Decreto 9 luglio 2012 Disciplina dell'impiego negli integratori alimentari di sostanze e preparati vegetali (G.U. 21-7-2012 serie generale n. 169).
Italy: Ministerio Della Salute Estratti Vegetali Non Amessi Negli Integratori Alimentari.
Netherlands: Nieuwe Voedsel en Waren Autoriteit, Ministerie van Landbouw, Natuur en Woedselkwaliteit (2010). Warenwetbesluit Kruidenpreparaten.
Austria: Federal Ministry of Health: Austrian Food Code, IVth ed., Section B31, Tea and tea-like products.
Austria: Federal Ministry of Health: Plants and plant parts recommended for use without dose restrictions.
Austria: Federal Ministry of Health: Plants and plant parts not recommended for use.
Sweden: Livsmedelsverket: Förteckning över växter och växtdelar som är olämpliga i livsmedel (VOLM), latest update May 2010.
Slovakia: Výnos Ministerstva pôdohospodárstva Slovenskej republiky a Ministerstva zdravotníctva Slovenskej republiky zo 6. apríla 2005 č. 2089/2005-100, ktorým sa vydáva hlava Potravinového kódexu Slovenskej republiky upravujúca pochutiny.
United Kingdom: Medicines and Healthcare Products Regulatory Agency (MHRA): Prohibited or restricted herbal ingredients, http://www.mhra.gov.uk/Howweregulate/Medicines/Herbalmedicinesregulation/Prohibitedorrestrictedherbalingredients/index.htm.
United Kingdom: Medicines and Healthcare Products Regulatory Agency (MHRA): List of herbal ingredients and their reported uses, http://www.mhra.gov.uk/home/groups/is-pol/documents/websiteresources/con009277.pdf.